Intelligent Security Management and Control in the IoT

SCIENCES

Networks and Communications, Field Director – Guy Pujolle

Network Management and Control, Subject Head – Francine Krief

Intelligent Security Management and Control in the IoT

Coordinated by
Mohamed-Aymen Chalouf

iSTE

WILEY

First published 2022 in Great Britain and the United States by ISTE Ltd and John Wiley & Sons, Inc.

ISTE Ltd
27-37 St George's Road
London SW19 4EU
UK

www.iste.co.uk

John Wiley & Sons, Inc.
111 River Street
Hoboken, NJ 07030
USA

www.wiley.com

Library of Congress Control Number: 2022931274

British Library Cataloguing-in-Publication Data
A CIP record for this book is available from the British Library
ISBN 978-1-78945-053-8

ERC code:
PE7 Systems and Communication Engineering
PE7_1 Control engineering
PE7_8 Networks (communication networks, sensor networks, networks of robots, etc.)

Contents

1

Multicriteria Selection of Transmission Parameters in the IoT

Sinda BOUSSEN[1], Mohamed-Aymen CHALOUF[2] and Francine KRIEF[3]

[1] *Mediatron, University of Carthage, Tunis, Tunisia*

[2] *IRISA, University of Rennes 1, Lannion, France*

[3] *LaBRI, Bordeaux INP, Talence, France*

1.1. Introduction

Cognitive radio networks (CRN) and the Internet of Things (IoT) are concepts that are taking on more and more importance in modern communication systems. The IoT extends connectivity to the real world (Atzori *et al.* 2010; Perera *et al.* 2014, 2015; Xu e*t al.* 2014; Al-Fuqaha *et al.* 2015), by allowing an object (a sensor, smartphone, car, etc.) to interact with the existing internet infrastructure, to communicate with other objects and to collect and exchange data. An object can have several interfaces and so detect different access networks. When several access networks are available, it will be necessary to make a decision about selecting the access network best adapted to the current situation. However, with the increase in the number of connected objects, the spectrum is becoming a precious resource, one threatened with scarcity. Faced with this problem, we can imagine objects with cognitive capacities (Vlacheas *et al.* 2013). This will make it possible to optimize use of unoccupied radio frequencies while still minimizing interferences with priority users. In an intelligent radio environment (Haykin 2005; Akyildiz *et al.* 2006), we distinguish two types of users: priority users, called primary users, who have an exclusive right over some of the spectrum, and secondary users, also called

Intelligent Security Management and Control in the IoT,
coordinated by Mohamed-Aymen CHALOUF.

cognitive users, who have "opportunistic" access to the spectrum. The main functions of intelligent radio are spectrum detection, decision-making, spectrum sharing and spectrum mobility. There is a great deal of research focusing on IoT architectures based on intelligent radio (Shah *et al.* 2013; Vlacheas *et al.* 2013; Aijaz and Aghvami 2015; Khan *et al.* 2016), which has demonstrated the need to integrate intelligent radio into the IoT (Wu *et al.* 2014a).

In this chapter, we tackle the question of making a decision for effective access to an access network or radio channel in the IoT. To communicate, an object that has several interfaces and/or cognitive capacities can detect numerous access networks or radio channels. In this case, it is necessary to choose the access network or radio channel best adapted to the communication in question. Among those aspects that might be considered in this decision, we might find Quality of Service (QoS) constraints on the IoT application and the object's own energy constraints (Zhu *et al.* 2015; Shaikh *et al.* 2017). Thus, the decision-making module should be multicriteria and consider the dynamic context of the radio environment, the needs of the application in terms of QoS and the object's own energy constraints.

The remainder of the chapter is organized as follows. Sections 1.2 and 1.3 give an overview of recent work on managing vertical handover and spectrum handoff, respectively, in an IoT context. In section 1.4, we suggest a multicriteria decision-making module that makes it possible to select and adapt radio transmission perimeters in order to satisfy the QoS requirements of the transported data equally as well as the energy constraints of the communicating objects. The suggested module will be used equally as much where heterogenous networks co-exist as in the context of intelligent radio. The choice of access network (presence of several heterogenous networks) or radio channels (use of intelligent radio) depends on the number of types of contextual information such as the networks state, user preferences, the application constraints and the characteristics of the object. To evaluate the suggested decision-making module, we consider the domain of Vehicular Ad-hoc NETworks (VANETs) and examine two possible situations: with and without intelligent radio.

1.2. Changing access network in the IoT

Managing mobility, QoS and energy in heterogenous networks is a major challenge for modern communications systems. So, a number of offers have come to light in recent years to enable a user to stay connected as much as possible, in any place.

In cases where heterogenous networks co-exist, an object equipped with several interfaces should be able to choose the access network best adapted to its needs, from among those available. Vertical handover (VHD) decision-making algorithms were designed to this end. The main purpose of these algorithms is to ensure the required QoS while still guaranteeing a transparent mobility between the different access technologies. Recently, several solutions have been suggested to overcome these challenges in an IoT context. These solutions (VHD algorithms) can be classed into five categories (Kassar *et al.* 2008; Zekri *et al.* 2012; Bhute *et al.* 2014) depending on the approach considered: classical, artificial intelligence (IA), cost-dependence, multiple attribute (Xiao and Li 2018) and decision-making, depending on the context.

In addition to the constraint linked to QoS and to mobility, the objects have an energy constraint. It is therefore important to consider energy consumption when selecting the most appropriate access network. Thus, several VHD solutions focus on energy efficient networks (Tuysuz and Trestian 2017). Most of these solutions used IEEE 802.21 MIH (Iqbal *et al.* 2019) and ANDSF (Access Network Discovery and Selection Function) protocols to collect information. Moreover, existing VHD systems make it possible to save energy by minimizing the scanning/detection time needed to discover the wireless network or to select the access point most economical with energy (Xenakis *et al.* 2011).

1.3. Spectrum handoff in the IoT

To avoid the scarcity of frequencies introduced above, the IoT can use the concept of intelligent radio. Many researchers have studied problems linked to spectrum management and channel allocation (Wu *et al.* 2014b; Koushik *et al.* 2018; Tarek *et al.* 2020). In Kumar *et al.* (2016), the authors studied spectrum handoff schemas in a CRN context and identified three types of transfer. First of all, the reactive approach, where the spectrum handoff and reconfiguration of the radio frequency happen after the primary user has been detected. Then, there is the proactive approach, where the spectrum handoff and frequency reconfiguration take place before a primary user has occupied the channel. Finally, the hybrid approach combines the two previous approaches. Recently, intelligent radio has been used in the IoT, especially in the two vehicular ad-hoc networks. This use has made it possible to improve the network's spectrum efficiency and the experience of itinerant users by optimizing vehicle communication capacities and the QoS of applications such as road safety and traveler entertainment (Singh *et al.* 2014; Kumar *et al.* 2017).

To ensure continuity of service, we have opted for a proactive approach using a prediction module tasked with calculating the probability of future channel availability as well as the average time it is available.

1.4. Multicriteria decision-making module for an effective spectrum handoff in the IoT

In the IoT, the multicriteria solution seems most appropriate for making decisions about VHD, and by considering a great deal of contextual information to ensure effective selection of an access network/radio channel. To overcome the complexity of implementation, we opt for a cost function solution that calculates different candidates' scores (access network or radio channel) and selects the one with the highest score. In the approach retained, we focus both on the QoS and on energy consumption, with the weight attributed to these parameters depending on the general communication context (network, application, object and user). Contextual information from the network is collected with the help of the surveillance module (context of multiple access networks) and of the intelligent radio-detection module (CRN context).

1.4.1. *General architecture*

In this section, we detail the general architecture we propose (Figure 1.1). In the approach we have retained, based on costs, the decision-making mechanism makes it possible to select the access network best adapted in the case of an object with several network interfaces and, on the other hand, to adapt the transmission parameters (channel, frequency, modulation, etc.) in the case of an intelligent radio network context. This selection will be based on the information available about the application, the user, the current radio conditions and predictions about how they will evolve. Information on the radio conditions is provided by the radio channel detection module (Akyildiz *et al.* 2006) in the CRN context and by the network monitoring module in the context of Multiple Radio Access Networks (M-RAN). The multicriteria decision-making module and the other general architecture modules are implemented within the object, which may be a car in the case of vehicular networks.

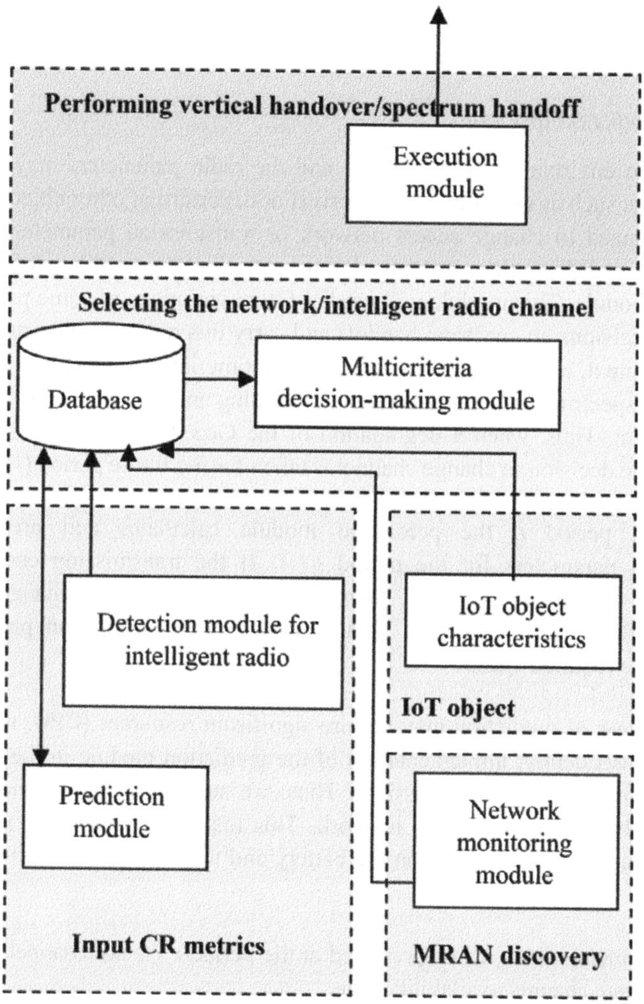

Figure 1.1. *Proposed architecture for a context aware IoT device/object. For a color version of this figure, see www.iste.co.uk/chalouf/intelligent.zip*

1.4.1.1. *Detection for an intelligent radio module*

In the context of CRNs, the detection module provides information on the accessible radio channels, their quality and their rates of occupation. A candidate radio channel (noted in the CR channel suite) is characterized by a central (fixed) frequency, a passband width (fixed), modulations and schemas for possible codings (fixed), a probability of availability (predicted), an average availability time (predicted), an energy cost (estimated), a packet loss rate (estimated), a passband (fixed), an average delay (estimated) and an average jitter (estimated).

1.4.1.2. *Prediction module*

The radio environment is not stable and the radio parameters may change for many reasons such as mobility and the arrival or departure of other objects. Thus, an object may need to change access network or transmission parameter values. To avoid delays and disruptions to transmission, unusable spectrum handoffs, we use a prediction module. This module will predict future variations of some parameters to anticipate decisions on spectrum handoff and carry this out at an opportune time. In the CRN context, a period could be equal to the time interval allocated to an object to carry out spectrum detection operations, signaling and the transfer of data along a given channel. Thus, when a degradation of the QoS is perceived for the current period (t), the decision to change channel is taken for the future period ($t + 1$).

At each period t, the prediction module calculates and predicts some transmission parameters for the period $t + 1$. If the transmission conditions are satisfactory for the period $t + 1$, then it is not necessary to change channels. Otherwise, the object considered should modify the transmission parameters to guarantee the required QoS.

The process of prediction may require significant resources (CPU, memory and energy). Consequently, implementation of the prediction module in the object may be very restrictive, indeed impossible. Thus, we suggest migrating the prediction process to the Fog radio access network. This makes it possible to improve the object's capacity, to economize on the battery and to provide the user with a better experience.

The parameters that will be predicted at the period $t + 1$ are channel availability and the average channel availability time.

1.4.1.2.1. Probability of channel availability at period t + 1

In a CRN context, according to the analytical model defined in Song and Xie (2012), the predicted probability that a candidate channel i is inactive at time t can be expressed as follows (equation [1.1]):

$$\Pr(N_i(t) = 0) \text{ if } \begin{cases} T_i^k + L_i^k < t \\ T_i^{k+1} \geq t, \end{cases} \text{ and } T_i^k \geq t, \quad \begin{matrix} k \geq 1 \\ k = 0 \end{matrix} \qquad [1.1]$$

where T_i^k represents the arrival time of the kth packet and designates the length of the kth PU packet (data from one primary user) on the channel i. $N_i(t)$ represents the state of the channel, which is a random binary variable with values 0 and 1 representing, respectively, the inactive and the occupied states.

The probability that the channel i is in active at an instant t can be obtained by equation [1.2]:

$$\Pr(N_i(t)=0)=\int_0^\infty\left[\sum_{k=1}^{\infty}\Pr(T_i^k+L_i<t|k)\Pr(T_i^{k+1}\geq t|k)\Pr(k)+\Pr(T_i^1\geq t)\Pr(k=0)\right]f_{L_i}(l)\mathrm{d}l$$

$$=\int_0^\infty\left\{\sum_{k=1}^{\infty}\left[\frac{(\bar{\lambda}_i(t-L_i))^k}{k!}\mathrm{e}^{-\bar{\lambda}_i(t-L_i)}\right]\left(\frac{(\bar{\lambda}_i t)^k}{k!}\mathrm{e}^{-\bar{\lambda}_i t}\right)\frac{(\bar{\lambda}_i t)^k}{k!}\mathrm{e}^{-\bar{\lambda}_i t}+\mathrm{e}^{-2\bar{\lambda}_i t}\right\}f_{L_i}(l)\mathrm{d}l \quad [1.2]$$

where X_i^k represents the interarrival of the PU packets. Supposing that the PU packets arrive following a Poisson process (Wang and Wang 2008), distributed exponentially with the average packet arrival rate per second λ_i, the length of the PU packet following $fLi\ (l)$.

Based on the prediction above, the candidate channel i is considered inactive for the following time interval when Pr $(N_i\ (t)\ =\ 0)\geq \tau_H$, where τ_H is the probability threshold for a channel to be considered inactive at the end of the current frame.

1.4.1.2.2. Average channel availability time

In an intelligent radio context, we used the model suggested by Song and Xie (2012) to predict the average channel availability time at instant t. The OFF period represents the period the channel is unavailable. For the ith channel, the cumulative distribution function (CDF) over the duration of the OFF period is given by equation [1.3]:

$$\Pr(t_{\text{off}}<x)=\int_0^\infty\int_0^{l+x}\bar{\lambda}_i\mathrm{e}^{-\bar{\lambda}_i t}f_{L_i}(l)\mathrm{d}t\mathrm{d}l$$

$$=\int_0^\infty\left(1-e^{-\bar{\lambda}_i(l+x)}\right)f_{L_i}(l)\mathrm{d}l, \quad [1.3]$$

where t_{off} represents the duration of the OFF period.

In order to support at least one SU packet (data from a secondary user), the probability that the inactive duration of the *i*th channel is longer than a transmission time of the frame η should be greater than or equal to θ (see equation [1.4]):

$$\Pr(tj,\ \mathrm{OFF}>\eta) \geq \theta, \tag{1.4}$$

where θ is the probability threshold for a channel to be considered inactive for the next packet transmission.

1.4.1.3. *Object characteristics*

An object is characterized by the emission power P_{Tx} and state of its battery. The battery state or its remaining life (expressed in hours) can be calculated (Grace *et al.* 2009) on the basis of the battery's capacity, its tension, its emission power and the constant K, which represents the power required for signal treatment.

1.4.1.4. *Network monitoring module*

In the M-RAN context, the network monitoring module provides information on the available access networks. These available access networks have different characteristics in terms of bandwidth, delay, loss rate, channel occupation rate and energy expenditure. The information collected (Carneiro *et al.* 2009) is analyzed to take a specific decision based on the services required.

1.4.1.5. *Database*

The input metrics needed by the suggested multicriteria decision-making module include the characteristics of the access network or radio channel, the characteristics of the object, application requirements, users' QoS preferences and the energy constraints and user sensitivity to some bands that could interfere with medical equipment in an e-health context, for example. All this information is stored in the databases. We distinguish types of input parameters: the measured, predicted and static parameters. The detection module for the intelligent radio and the monitoring module in the M-RAN context provide measured parameters. The prediction module will be responsible for predicting channel availability and the average duration for which they are available. The static parameters include fixed characteristics of the network or radio channel, user preferences and application requirements. The application needs we retain are linked to the QoS required by the application, such as the minimal bandwidth, the maximum delay allowed, the maximum jitter tolerated *m* and the maximum packet loss rate tolerated.

1.4.1.6. *Multicriteria decision-making module*

Depending on the needs of the application (data type, transmission urgency), the spectrum availability and the remaining battery life, the multicriteria decision-making module will select the most appropriate access network or radio channel to ensure transmission. This module considers both the energy constraints and the QoS and will be detailed in section 1.4.2.

To ensure decision stability, which is an important criteria (Wang *et al.* 2014), and to avoid inefficient or unhelpful transfer decisions in both the contexts considered (M-RAN or CRN), we suggest using a waiting time (timer) between the selection and execution stages. At the end of this time, the device checks that the following conditions are still valid: (1) the network/candidate channel is still detected by the object, and (2) this candidate's score (network/channel) is still higher than that of the current network/channel. If these conditions are verified, then the transfer (network or channel) is run, otherwise it is canceled and the selection process restarts from the beginning.

As this wait can be an obstacle to continuity of service, a compromise needs to be found. So, we suggest adjusting this waiting time depending on the context of the decision.

1.4.1.7. *Module for running vertical handover/spectrum handoff*

The running stage refers to establishing the connection with the new access network (M-RAN context) or changing channel/transmission parameters (CRN context). In the M-RAN context, three cases are possible so long as the old link is released before (Hard Handover), during (Seamless Handover) or after (Soft Handover) the new link is established.

1.4.2. *Decision-making flowchart*

The decision-making module, detailed in section 1.4.1, makes it possible to select the most appropriate access network (M-RAN) or the radio channel (CRN). This selection depends on many contextual parameters, which include characteristics of the radio environment, application needs, object capacities, etc. In our case, selecting an access network or CR channel means selecting all the parameters that characterize transmission. In fact, an access network can be considered as one candidate channel among others. This is a key element for guaranteeing that our suggestion functions in all possible cases (as Figure 1.2 shows). The n candidate networks/channels detected are written $C1$, $C2$, $C3$, …, Cn. On each channel (Ci), it is possible to transmit with the help of m modulations M_1, M_j, M_m. Each C_i channel

(Figure 1.2) is characterized by the probability of availability A_i, the occupation rate O_i, the bandwidth, the delay, the loss and the energy cost. Each channel is also characterized by one or more transmission models ($MOD_{i,m}$, where i designates the channel and m designates the modulation and coding).

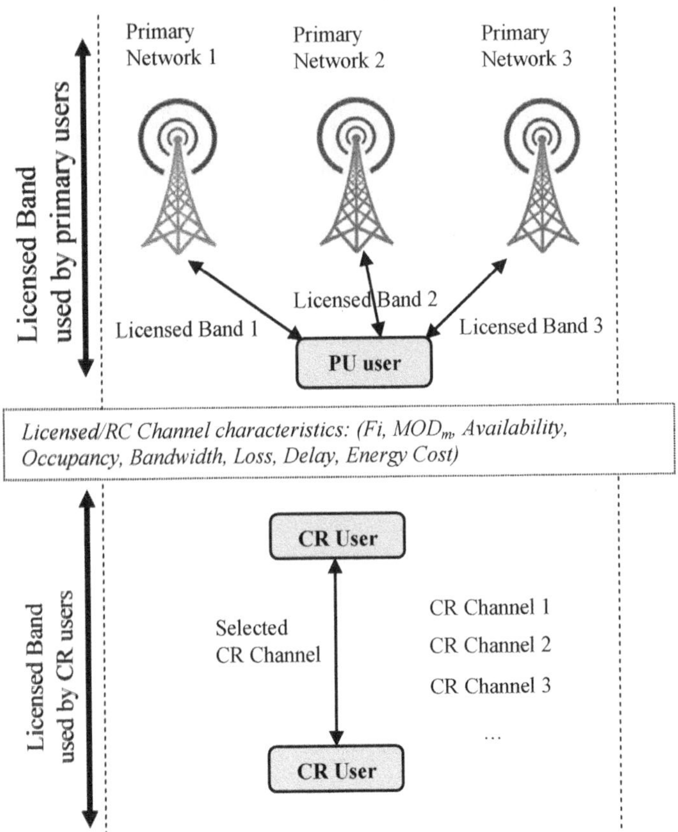

Figure 1.2. *General context of decision-making*

The suggested multicriteria decision-making module is based on the concept of utility to reduce the complexity of such a decision. The utility function used to calculate the scores is $U(x)=1-e^{-\alpha x}$ with x the parameter of the decision vector and α its corresponding weight ($\alpha > 0$). For $x \geq 0$, $U(x) \in [0, 1]$.

The suggested selection algorithm is formed of the following stages. The first stage consists of verifying the availability of the different channels detected to eliminate poor candidates that have an availability probability below a defined threshold or whose availability time is less than a given duration. This preselection makes it possible to avoid unhelpful score calculation. At the second stage, the channels selected during the first stage are classed according to their energy scores. At the third stage, the decision-making module calculates the QoS score for each selected channel. After having calculated scores linked to QoS and to energy, the total score is calculated for each channel selected depending on the weight given to each aspect considered: QoS and energy. This is stage 4. The attribution of weight to criteria should reflect the importance of each decision-making criteria depending on the overall context (application, battery, etc.). Finally, the selected channel will be the candidate with the highest total score.

1.4.2.1. *Stage 1: selecting available access networks/CR channels*

In the M-RAN context, this processing stage is very simple. In fact, each access network detected is available. However, in a CRN context, this stage requires the following processing. First of all, it involves checking the availability of the current channel k and determining the probability $\Pr(C_k(t) = 0)$ that the channel k will be inactive at instant t. The secondary user should switch to a new channel when the availability of the current channel is lower than the probability threshold (equation [1.5]):

$$\Pr\left(C_k(t)=0\right) \geq \theta, \qquad [1.5]$$

where $C_k(t)$ represents the current channel k, which is a binary random variable with the values 0 and 1 representing the inactive and the occupied states, respectively.

τ_L is the probability threshold below which a channel is considered occupied and the secondary user should be allocated a spectrum handoff (Song and Xie 2012), that is, the current channel is no longer considered inactive at the end of transmission from the frame.

Then, for each channel i detected, it remains to determine the probability of inactivity and the average availability of the channel for the next period.

The candidate channel i is considered inactive for the next period when $\Pr(C_i(t) = 0) \geq \tau_H$, where τ_H is the probability threshold for a channel to be considered inactive at the end of the current frame (Song and Xie 2012). Thus, the preselected channels are those that have a probability $\Pr\left(C_i(t)=0\right) \geq \tau_H$ and an average duration of availability higher than a threshold $\Pr\left(t_i\text{OFF} > \eta\ C_i(t)=0\right) \geq \theta$; t_iOFF represents the duration of the OFF period of the channel i.

To support at least one SU frame, the probability that the activity duration of the ith channel will be longer than one transmission time of the frame η should be higher than or equal to θ. θ is the probability threshold for a channel to be considered inactive for the next frame transmission (Song and Xie 2012).

Thresholds τ_L and τ_H are not fixed arbitrarily. The choice of these thresholds can have an impact on the number of channels filtered during this stage. For this, we will choose both these thresholds depending on the radio context, again, such as the noise that characterizes this environment or the inactivity of primary users. For example, when there are very few channels considered available, we can lower the value of $\tau_L = \tau_H$. On the contrary, if these channels are very numerous, the value of $\tau_L = \tau_H$ will be increased further. This will make it possible not only to select the channels with a high probability of availability, but also to reduce the number of candidates for which the decision-making module should run the remainder of the processing (calculation of different scores and decisions). Thus, making the right choice from among these thresholds will directly impact the efficiency of the approach retained as well as its global cost (CPU, memory, energy, etc.). We note that in some cases, τ_L could be different from τ_H.

1.4.2.2. *Stage 2: classifying channels depending on the energy cost of the transmission models*

Our aim is to minimize energy consumption by choosing the channels that use least energy. To do this, we classify the candidate channels depending on their energy score.

On each C_i channel, it is possible to transmit with the help of m modulations M_m. $(MOD)_{im}$ designates the transmission models possible on each channel C_i, where i represents the channel and m designates the modulation schema and the coding. Because the transmission power of the object has an impact on energy consumption, we suggest evaluating the transmission cost of the transmission models per available channel $(MOD)_{im}$. This makes it possible to select the channel that most reduces energy consumption and so prolongs the battery life. In fact, the higher the transmission power, the better the signal-to-noise ratio, but a higher transmission power also means greater energy consumption (Krief 2012).

To select the most appropriate channel, the multicriteria decision-making module suggested should (1) calculate the minimum transmission power needed to transmit correctly on a given channel C_i with a modulation M_m, toward the other pair that could be an access point (M-RAN context) or another object (CRN context), and (2) eliminate the transmission models $(MOD)_{im}$ necessitating a minimal transmission power greater than the device's maximum transmission power.

To avoid rerunning the calculations at each execution of the selection algorithm, we suggest making calculations upstream and to store them within each object or on a remote device (Fog, Cloud, etc.). In fact, the minimum transmission power for each possible transmission model (channel, modulation) is calculated for different distances and stored on a Fog or Cloud device accessible to the objects.

The energy score is calculated according to the energy cost of a candidate channel and the battery life of the IoT device. For a given modulation and coding schema, the energy cost of a candidate channel is the average energy consumed per bit for the transmission on the channel. The energy cost of data transmission from the application j on this channel i (written $C_{i,j}$) is calculated depending on the band frequency, the uplink channel rate and the data flow from application (packet size and packet interarrival time). However, in the context of video streaming, other parameters can be considered to estimate this cost, such as video quality, the number of users and the quality of the channel (Zou *et al.* 2017).

The energy score $S_{\text{Energy}}(i, j)$ is given by equation [1.6]:

$$S_{\text{Energy}}(i,j) = \left(1 - e^{\left(-1/C_{i,j}\right)}\right) \cdot \left(C_{i,j} \otimes \text{EB}\right), \quad [1.6]$$

where $C_{i,j}$ represents the energy cost of transmission of data j from the application along the channel i and EB designates the object's battery level.

Finally, the selected channels are classed depending on their energy scores. These are calculated according to the energy cost of the transmission models retained $(\text{MOD})_{im}$ and according to the level of the battery of the IoT device.

1.4.2.3. *Stage 3: calculation of the QoS score of the transmission models*

In this stage, the multicriteria decision-making module should calculate the QoS score of the remaining candidate channels. The application's QoS score $S_{\text{QoS}}(i, j)$ makes it possible to estimate the QoS offered to the application j when the transmission uses $(\text{MOD})_{im}$ (channel i, modulation m).

$S_{\text{QoS}}(i, j)$ is based on the QoS parameters, that is the bandwidth, the loss rate, the delay and the jitter. The importance of these parameters depends on the application that is running. It is expressed by attributing weightings to different parameters.

The QoS $S_{\text{QoS}}(i, j)$ score is given by equation [1.7]:

$$S_{\text{QoS}}(i,j) = (4 - e^{-\alpha B.Bi} - e^{-\alpha D/Di} - e^{-\alpha G/Gi} - e^{-\alpha E/Ei}) \cdot (B_i \otimes b_j) \cdot (d_j \otimes D_i)$$

$$(g_j \otimes G_i) \cdot (e_j \otimes E_i) \cdot (v_i \otimes V) \cdot A_i, \qquad [1.7]$$

where

- B_i is the bandwidth available via the channel i;
- b_j is the minimal bandwidth required by the application j;
- E_i is the loss rate via the channel i;
- e_j is the maximum loss rate authorized by the application j;
- D_i is the average delay in the channel i;
- d_j is the maximum delay supported by the application j;
- G_i is the average jitter in the channel i;
- g_j is the maximum jitter supported by the application j;
- v_i is the maximum speed authorized on the channel i (v_i depends on the network coverage);
- V is the speed of the IoT object;
- α_B is the weight of the bandwidth;
- α_D is the weight of the delay;
- α_G is the weight of jitter;
- α_E is the weight of the loss rate;
- $x_i \otimes y_j = 1$ if $x_i > y_i$, otherwise 0;
- A_i is the availability probability of the channel at the period $t + 1$.

The formula used to calculate the application's QoS score j makes it possible to distinguish the different classes of service by adjusting the threshold values (b_j, d_j, g_i, e_j) and the weight criteria (α_B, α_D, α_G, α_E).

1.4.2.4. *Stage 4: attribution of weight and calculating the final score*

This last stage will be responsible for calculating the total score $S_T(i, j)$ for each candidate channel by weight the energy scores and the QoS. The weightings attributed to both scores will make it possible to select the IoT application domain (e.g. VANET and e-health).

This stage will therefore calculate the total score $S_T(i, j)$ for each candidate channel depending on the weight attributed to each aspect. The weight attribution for the criteria should reflect the importance of each decision criteria depending on the application context or, depending on the case, the preferences of the user ($0 \leq \delta \leq 1$).

The total score $S_T(i, j)$ is given by equation [1.8]:

$$S_T(i,j) = \delta \cdot S_{\text{QoS}}(i,j) + (1-\delta) \cdot S_{\text{Energy}}(i,j) \quad [1.8]$$

The selected channel will be the candidate with the highest total score. Finally, this information will be registered in the database for future use when a similar case occurs.

1.4.3. *Performances evaluation*

The multicriteria decision-making module that we suggest should be useful in different IoT contexts, particularly in M-RAN and CRN contexts. To illustrate the use of the retained approach in these two contexts and evaluate performances in each of them, we consider, in this section, two significant IoT use-cases: VANET and CR-VANET (Cognitive Radio VANET). In a VANET network, the On-Board Unit (OBU) is installed in the vehicle and includes a wireless transmitter receiver and various sensors, whereas a Road-Side Unit (RSU) is deployed in strategic locations along the route to facilitate communication between the vehicle and the infrastructure (V2I: Vehicle to Infrastructure). In V2I, information is exchanged between the vehicle and the RSU, or potentially a cellular network. VANET applications can be divided into three main categories (Javed *et al.* 2014): road safety (collision detection, cooperative driving, etc.), traffic management (route guidance, traffic light synchronization, etc.) and user infotainment (Web, streaming audio/video, etc.). Safety applications do not tolerate a transmission delay higher than 0.5 s (Javed *et al.* 2014), whereas traffic management applications are less demanding, with a tolerated latency between 0.1 and 1 s (Javed *et al.* 2014). As for infotainment applications, they generally accept greater latency, in order of 1–5 s. Nevertheless, some applications of this type, like multiplayer games, may require lower latencies of 0.1–1 s (Javed *et al.* 2014).

Based on the access networks detected or the channel available, and depending on the QoS requirements of the VANET applications, the suggested multicriteria decision-making module will select the best access network or radio channel available for a specific case. To do this, the utility function will calculate the scores of different candidate networks/radio channels to select the one that has the highest score.

Since we are considering combustion rather than electrical vehicles, we suppose that there is no energy constraint. Thus, the scores calculated from the different candidate channels are based only on the QoS parameters.

To evaluate the approach retained, we consider two representative types of transmission: multimedia emergency notification and infotainment applications, especially restaurant reservation. Table 1.1 summarizes the QoS requirements of these services.

Type of service	Relevance of QoS parameters
Multimedia emergency notification	Latency (+++), bandwidth (+++) and packet loss (+)
Restaurant reservation	Latency (+), bandwidth (+) and packet loss (+)

Table 1.1. *Some vehicular network services and the corresponding QoS parameters*

Table 1.2 shows the estimation of the weight of the data flow, the delay and the packet loss rate for services considered in the vehicle networks. These weightings are based on the importance of each parameter in Table 1.2.

Type of service	Weight estimation		
	Delay	BP	PLR
Multimedia emergency notification	0.43	0.43	0.14
Restaurant reservation	0.33	0.33	0.33

Table 1.2. *Weight estimation for services considered in vehicular networks*

1.4.3.1. *Scenario 1: network selection – the case of V2I communications*

In this first scenario, we consider urban VANETs where the scene of the accident is captured by the first vehicle on the scene. Then, this vehicle will transmit the video to the rescue teams so that they can manage the emergency more effectively. In such a situation, the total transmission delay for a multimedia message is limited to a few seconds to ensure notification in real time (Javed *et al.* 2014). For this type of traffic (Table 1.1), the delay and the flow are considered dominant attributes.

As Figure 1.3 shows, the network architecture in a city is based on the vehicles with OBUs and infrastructures such as RSUs, WiFi access points and LTE eNB base stations for the 4G cell phone network. We suppose that each vehicle is equipped with three radio interfaces: 4G interface, WiFi interface and 802.11p interface.

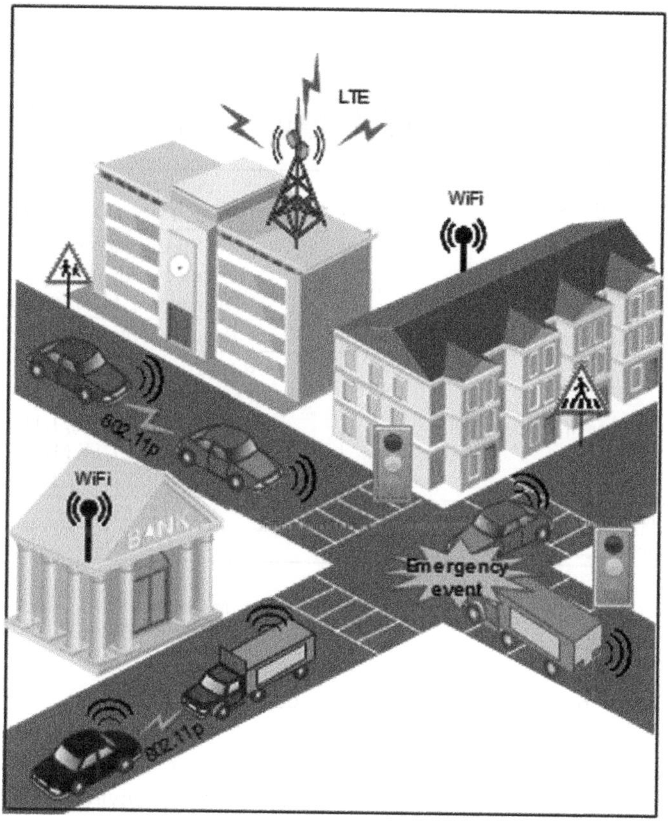

Figure 1.3. *Transmission of video from the scene of an accident in VANET networks*

The simulation of this scenario, under the network simulator ns3, involves a WiFi access point and a LTE base station offering a theoretical flow of 11 Mbps and 24 Mbps, respectively. At the start, the number of active users is 4 for the WiFi cell and 10 for the LTE cell. The simulation parameters are listed in Table 1.3.

The effective data flow observed by the user of an LTE or WiFi network may be much lower than the theoretical flows stated and defined by the norms. The main factors influencing the effective flow are the number of active users sharing the bandwidth within a cell, the bandwidth frequency allocated to the network operator and the distance between the terminal and the relay antenna. These factors will also influence other QoS parameters such as latency and the average loss rate. For these reasons, the suggested multicriteria decision-making module will calculate the scores of the different networks detected depending on the estimated available bandwidth, the average delay measured and the packet rate loss measured.

Parameters	Values
Duration of simulation	50 s
Traffic video	Send interval 1 ms Packet size 1,000 octets
Witness vehicle	Speed: stationary Distance – AP WiFi: 20 m Distance – LTE BS: 60 m
Mobility model	*Random Waypoint Model*
Number of active users	At t = 10 s, WiFi (4), LTE (10) At t = 20 s, WiFi (10), LTE (20) At t = 30 s, WiFi (20), LTE (10) At t = 40 s, WiFi (10), LTE (4) At t = 50 s, WiFi (10), LTE (10)
Theoretical flow for WiFi	11 Mbps
Theoretical flow for LTE	24 Mbps

Table 1.3. *Simulation parameters for scenario 1*

The number of active users may vary during the simulation (Table 1.3) as vehicles or pedestrians depart or arrive. The variation in this parameter influences the QoS parameters characterizing the candidate networks and, consequently, their scores. Figure 1.4 shows the effect of the dynamic network environment on optimal network selection for the service video. Depending on the configuration adopted for this scenario, the results of the simulation show that LTE is selected as an optimal network. In fact, when the LTE score increases because of the dynamic network environment, the optimal network passes from the WiFi to the LTE for the network selection decision. To ensure the stability of the decision and avoid the ping-pong effect, we wait for the following period to make the decision to be sure that the right decision is made.

Considering this example, if (1) the execution period of the suggested decision-making module is not adjusted automatically and (2) the transfer is made after the new trend is confirmed, the transfer will take place at 40 s. This is much later than the changeover time estimated at 23 s. This is why, as we said previously, it is very important to adjust the execution period of the suggested decision-making module automatically.

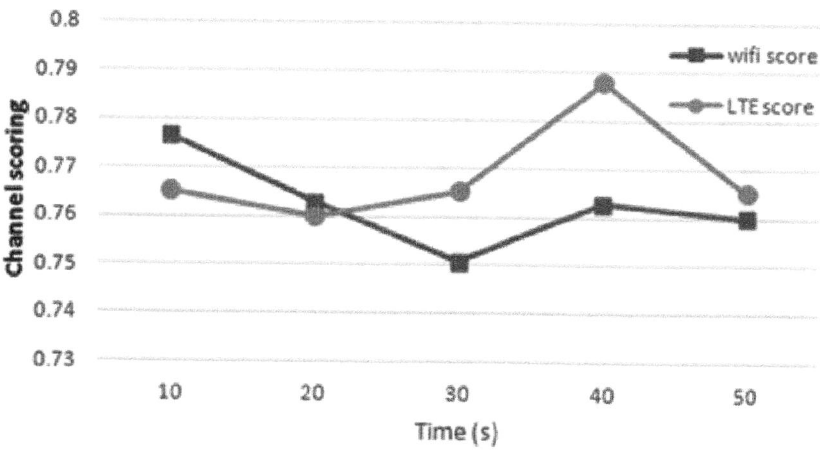

Figure 1.4. *Score variation in VANET networks for transmission of a video message*

1.4.3.2. *Scenario 2: selecting the radio channel – the case of CR-VANET*

Thanks to its cognitive capacity, the compatible intelligent radio vehicle can search for spectrum holes (white spaces) and take opportune decisions to transmit without interfering with the primary users. It can also make adaptations such as modifying the transmission power or the radio channels to meet QoS requirements for vehicle applications.

In this second scenario, we will consider a CR-VANET highway environment in the countryside and we will focus on infotainment applications, especially restaurant reservation. On the topology of CR-VANET highways, the speed of vehicles on the highways is significantly higher than in urban scenarios.

To evaluate performances, we consider the network architecture of a CR-VANET (Figure 1.5) where each vehicle is equipped with wireless communication interfaces (such as a 4G interface and a WiFi interface) and has intelligent radio capacities.

The simulations were made using ns3 paired with SUMO to generate real traffic mobility. The simulation parameters are listed in Table 1.4. For the same reasons cited in the previous scenario, the energy aspect will not be considered here.

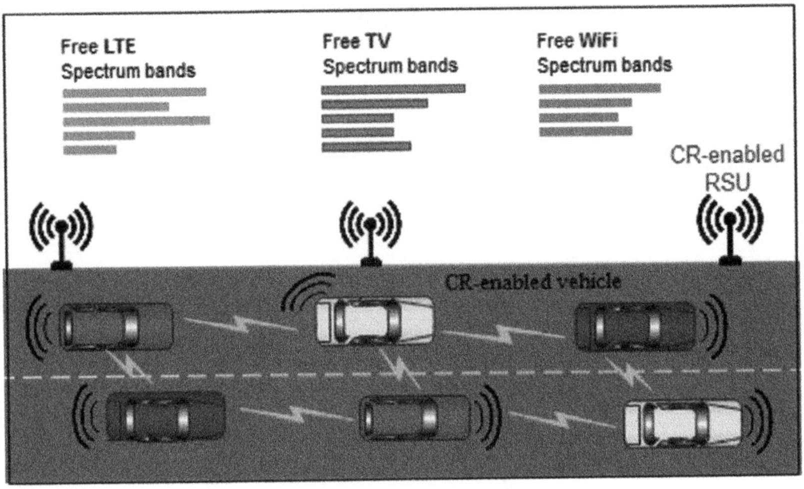

Figure 1.5. *Entertainment service on a highway in CR-VANET networks*

Parameters	Values
Simulation duration	50 s
Data traffic	Flow: 64 kbps
Vehicle mobility model	Car Following Model
Intelligent radio channels available	TV white space band WiFi white space band LTE white space band

Table 1.4. *Simulation parameters for scenario 2*

For the vehicle behavior model, we have adopted the *Car Following Model*, which is based on the following parameters: vehicle acceleration, vehicle deceleration, vehicle length, maximum speed of the vehicle and driver imperfection. The values of these parameters are presented in Table 1.5.

Based on the spectral resources detected, the suggested multicriteria decision-making module will calculate the scores of different channels and select the one with the highest score.

Parameters	Values
Vehicle acceleration	2.5 m/s^2
Vehicle deceleration	4.6 m/s^2
Average vehicle length	5 m
Maximum vehicle speed	140 km/h
Driver imperfection	0.5

Table 1.5. *Simulation parameters for the SUMO vehicle*

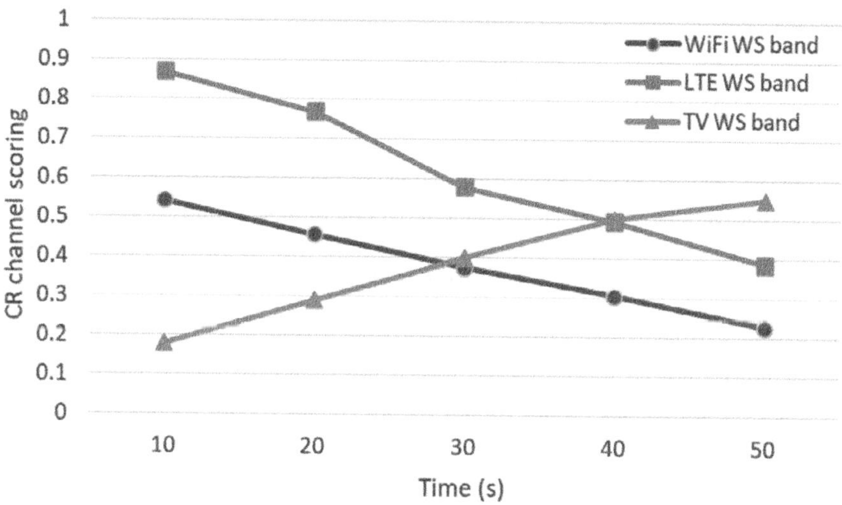

Figure 1.6. *Variation of the scores of intelligent radio channels for the entertainment service*

Figure 1.6 shows the impact of the dynamic network environment (such as the available bandwidth, the channel's availability probability and the speed of the vehicle) on selecting the channel best adapted to the infotainment service (restaurant reservation). In Figure 1.6, we can observe that with the high mobility of candidate vehicles on the highways, the channel corresponding to the WiFi bandwidth is not a good candidate. We also observe a degradation of the score of the radio channel corresponding to the bandwidth of the LTE and an increase in the score of the channel corresponding to the TV bandwidth. The evolution of the scores over time justifies the decision to carry out a spectrum handoff, from the suggested module to the TVWS channel. Before running the spectrum handoff, we wait for the choice of the best channel to be confirmed for the following period. This will enable us to avoid unhelpful transfers. And to avoid running the spectrum handoff very late, this wait time will be adjusted automatically. In this scenario, we observe that the choice of TVWS is fully adapted to this context, not only in terms of QoS but also from the perspective of mobility. Indeed, this spectral band is characterized by a long range that is best suited to scenarios with high mobility.

1.5. Conclusion

In this chapter, we tackled the question of decision-making for effective access to a radio network or a spectrum band in the IoT. An IoT object, having several interfaces and/or with cognitive capacities, can detect several access networks or radio communication channels. Thus, it may be led to choose the access network or radio communication channel that best meets the QoS constraints of the IoT application as well as its energy constraints. Selecting the most appropriate network or radio communication channel will allow the object to remain best connected. In this chapter, we focused on the functioning of the multicriteria decision-making module that we have suggested to tackle the problem of scalability. However, many approaches (Lounis *et al.* 2012; Gia *et al.* 2015; Guo *et al.* 2017; Firouzi *et al.* 2018; Shrestha *et al.* 2018; Khan and Lee 2019) have been proposed to solve the scalability issues in an IoT system. In our context, this question remains very important. This is why we plan to study it in future work.

1.6. References

Aijaz, A. and Aghvami, A.H. (2015). Cognitive machine-to-machine communications for Internet-of-Things: A protocol stack perspective. *IEEE Internet of Things Journal*, 2(2), 103–112.

Akyildiz, I.F., Lee, W.-Y., Vuran, M.C., Mohanty, S. (2006). NeXt generation/dynamic spectrum access/cognitive radio wireless networks: A survey. *Computer Networks Journal*, 50(13), 2127–2159.

Al-Fuqaha, A., Guizani, M., Mohammadi, M., Aledhari, M., Ayyash, M. (2015). Internet of Things: A survey on enabling technologies, protocols and applications. *IEEE Communications Surveys & Tutorials*, 17(4), 2347–2376.

Atzori, L., Iera, A., Morabito, G. (2010). The Internet of Things: A survey. *Computer Networks*, 54(15), 2787–2805.

Bhute, H.A., Karde, P.P., Thakare, V.M. (2014). Vertical handover decision strategies in heterogeneous wireless networks. *Proceedings of International Conference on Recent Trends in Information, Telecommunication and Computing, ITC 2014.* doi:02.ITC.2014.5.71.

Carneiro, P., Fortuna, P., Ricardo, M. (2009). Flow monitor: A network monitoring framework for the network simulator 3 (NS-3). *Proceedings of the 4th International ICST Conference on Performance Evaluation Methodologies and Tools (VALUETOOLS '09).* doi:10.4108/ICST.VALUETOOLS2009.7493.

Firouzi, F., Farahani, B., Ibrahim, M., Chakrabarty, K. (2018). Keynote paper: From EDA to IoTeHealth: Promises, challenges, and solutions. In *IEEE Transactions on Computer-Aided Design of Integrated Circuits and Systems*. doi:10.1109/tcad.2018.2801227.

Gia, T.N., Rahmani, A., Westerlund, T., Liljeberg, P., Tenhunen, H. (2015). Fault tolerant and scalable IoT-based architecture for health monitoring. In *IEEE Sensors Applications Symposium (SAS)*. IEEE, Zadar.

Grace, D., Chen, J., Jiang, T., Mitchell, P.D. (2009). Using cognitive radio to deliver "green" communications. In *2009 4th International Conference on Cognitive Radio Oriented Wireless Networks and Communications*. IEEE, Hanover.

Guo, H., Ren, J., Zhang, D., Zhang, Y., Hu, J. (2017). A scalable and manageable IoT architecture based on transparent computing. *Journal of Parallel and Distributed Computing*, 118(1) 5–13. doi:10.1016/j.jpdc.2017.07.003.

Haykin, S. (2005). Cognitive radio: Brain-empowered wireless communications. *IEEE Journal on Selected Areas in Communications*, 23(2), 201–220.

Iqbal, J., Khan, M., Afaq, M., Ali, A. (2019). Performance analysis of vertical handover techniques based on IEEE 802.21: Media independent handover standard. *Transactions on Emerging Telecommunications Technologies*. doi:10.1002/ett.3695.

Javed, M.A., Ngo, D.T., Khan, J.Y. (2014). Multimedia communication for emergency services in cooperative vehicular ad hoc networks. In *Multimedia over Cognitive Radio Networks*, Hu, F., Kumar, S. (eds). CRC Press, Boca Raton.

Kassar, M., Kervella, B., Pujolle, G. (2008). An overview of vertical handover decision strategies in heterogeneous wireless networks. *Journal of Computer Communications*, 37(10), 2607–2620.

Khan, U.A. and Lee, S.S. (2019). Multi-layer problems and solutions in VANETs: A review. *Electronics 2019*, 8, 204.

Khan, A.A., Rehmani, M.H., Rachedi, A. (2016). When cognitive radio meets the Internet of Things? In *2016 International Wireless Communications and Mobile Computing Conference (IWCMC)*. IEEE, Paphos.

Koushik, A.M., Hu, F., Kumar, S. (2018). Intelligent spectrum management based on transfer actor-critic learning for rateless transmissions in cognitive radio networks. *IEEE Transactions on Mobile Computing*, 17(5), 1204–1215.

Krief, F. (ed.) (2012). *Green Networking*. ISTE Ltd, London, Wiley, New York.

Kumar, K., Prakash, A., Tripathi, R. (2016). Spectrum handoff in cognitive radio networks: A classification and comprehensive survey. *Journal of Network and Computer Applications*, 61, 161–188.

Kumar, K., Prakash, A., Tripathi, R. (2017). A spectrum handoff scheme for optimal network selection in NEMO based cognitive radio vehicular networks. *Wireless Communications and Mobile Computing*, 1–16.

Lounis, A., Hadjidj, A., Bouabdallah, A., Challal, Y. (2012). Secure and scalable cloud-based architecture for e-health wireless sensor networks. In *International Conference on Computer Communication Networks (ICCCN)*. IEEE, Munich.

Perera, C., Liu, C.H., Jayawardena, S., Chen, M. (2014). A survey on Internet of Things from industrial market perspective. *IEEE Access*, 2, 1660–1679.

Perera, C., Liu, C.H., Jayawardena, S. (2015). The emerging Internet of Things market-place from an industrial perspective: A survey. *IEEE Transactions on Emerging Topics in Computing*, 3(4), 585–598.

Shah, M.A., Zhang, S., Maple, C. (2013). Cognitive radio networks for Internet of Things: Applications, challenges and future. In *19th International Conference on Automation & Computing*. IEEE, London.

Shaikh, F.K., Zeadally, S., Exposito, E. (2017). Enabling technologies for green Internet of Things. *IEEE Systems Journal*, 11(2), 983–994.

Shrestha, R., Bajracharya, R., Nam, S.Y. (2018). Challenges of future VANET and cloud-based approaches. *Wireless Communications and Mobile Computing*. doi:10.1155/2018/5603518.

Singh, K.D., Rawat, P., Bonnin, J. (2014). Cognitive radio for vehicular ad hoc networks (CR-VANETs): Approaches and challenges. *Journal on Wireless Communications and Networking*, 49(2014). doi:10.1186/1687-1499-2014-49.

Song, Y. and Xie, J. (2012). ProSpect: A proactive spectrum handoff framework for cognitive radio ad hoc networks without common control channel. *IEEE Transactions on Mobile Computing*, 11(7), 1127–1139.

Tarek, A., Benslimane, M., Darwish, M. (2020). Survey on spectrum sharing/allocation for cognitive radio networks Internet of Things. *Egyptian Informatics Journal*. doi:10.1016/j.eij.2020.02.003.

Tuysuz, M.F. and Trestian, R. (2017). Energy-efficient vertical handover parameters, classification and solutions over wireless heterogeneous networks: A comprehensive survey. *Wireless Personal Communications*, 97, 1155.

Vlacheas, P., Giaffreda, R., Stavroulaki, V., Kelaidonis, D., Foteinos, V., Poulios, G., Demestichas, P., Somov, A., Biswas, A.R., Moessner, K. (2013). Enabling smart cities through a cognitive management framework for the Internet of Things. *IEEE Communications Magazine*, 51(6), 102–111.

Wang, L.-C. and Wang, C.-W. (2008). Spectrum handoff for cognitive radio networks: Reactive-sensing or proactive-sensing? In *Proceedings of IEEE International in Performance, Computing and Communications Conference (IPCCC 2008)*, Austin, TX, 343–348. 10.1109/PCCC.2008.4745128.

Wang, Y.-H., Huang, G.-R., Tung, Y.-C. (2014). A handover prediction mechanism based on LTE-A UE history information. In *International Conference on Computer, Information and Telecommunication Systems (CITS)*. IEEE, Jeju.

Wu, Q., Ding, G., Xu, Y., Feng, S., Du, Z., Wang, J., Long, K. (2014a). Cognitive Internet of Things: A new paradigm beyond connection. *IEEE Internet of Things Journal*, 1(2), 129–143. doi:10.1109/JIOT.2014.2311513.

Wu, Y., Hu, F., Kumar, S., Zhu, Y., Talari, A., Rahnavard, N., Matyjas, J.D. (2014b). A learning-based QoE-driven spectrum handoff scheme for multimedia transmissions over cognitive radio networks. *IEEE Journal on Selected Areas in Communications*, 32(11), 2134–2148.

Xenakis, D., Passas, N., Di Gregorio, L., Verikoukis, C. (2011). A context aware vertical handover framework towards energy-efficiency. In *Proceedings of the IEEE Vehicular Technology Conference (VTC)*.

Xiao, K. and Li, C. (2018). Vertical handoff decision algorithm for heterogeneous wireless networks based on entropy and improved TOPSIS. In *2018 IEEE 18th International Conference on Communication Technology (ICCT)*. IEEE, Chongqing.

Xu, L.D., He, W., Li, S. (2014). Internet of Things in industries: A survey. *IEEE Transactions on Industrial Informatics*, 10(4), 2233–2243.

Zekri, M., Jouaber, B., Zeghlache, D. (2012). A review on mobility management and vertical handover solutions over heterogeneous wireless networks. *Computer Communications*, 35, 2055–2068.

Zhu, C., Leung, V.C., Shu, L., Ngai, E.C.H. (2015). Green IoT for smart world. *Special Section on Challenges for Smart Worlds*. doi:10.1109/ACCESS.2015.2497312.

Zou, L., Javed, A., Muntean, G. (2017). Smart mobile device power consumption measurement for video streaming in wireless environments: WiFi vs. LTE. In *2017 IEEE International Symposium on Broadband Multimedia Systems and Broadcasting (BMSB)*. IEEE, Cagliari.

2

Using Reinforcement Learning to Manage Massive Access in NB-IoT Networks

Yassine HADJADJ-AOUL and Soraya AIT-CHELLOUCHE
Inria, CNRS, IRISA, University of Rennes 1, France

2.1. Introduction

Communications between objects in the Internet of Things (IoT) and particularly machine-to-machine (M2M) communications are considered as one of the most important evolutions of the Internet. Supporting these devices is, however, one of the most significant challenges that network operators need to overcome (Lin *et al.* 2016). In fact, the considerable number of these devices that might attempt to access a network at the same time could lead to heavy congestion, or even a total saturation, with all the consequences this might cause. In fact, as can be seen in Figure 2.1, a very limited number of devices trying simultaneously to access the network can reduce the network's performance to zero, independently of the access opportunities available (Bouzouita *et al.* 2016). In these circumstances, it seems clear that effective access control mechanisms are needed to maintain a reasonable number of access attempts.

The Third Generation Partnership Project (3GPP) group has identified the overload on the Random Access Network (RAN) as a priority at a very early stage and has suggested several solutions. Among the approaches suggested, the Access Class Barring (ACB), suggested in version 8, and its extension, the Extended Access Barring (EAB), suggested in version 11, are certainly the most effective strategies

Intelligent Security Management and Control in the IoT,
coordinated by Mohamed-Aymen CHALOUF.

(3GPP 2011). In fact, these approaches tackle the problem at its roots by preventing any attempts to access the network. However, these approaches merely provide a framework for congestion management, without providing a ready-made solution.

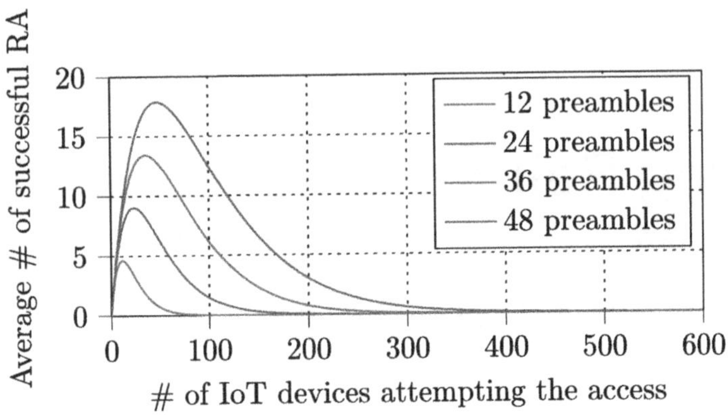

Figure 2.1. *Impact on performance of the number of IoT devices simultaneously attempting access*

The idea introduced in this chapter is quite simple, because it involves calculating a blocking factor. Nevertheless, a good implementation would require a good knowledge of the number of terminals ready to attempt access, so as to deduce from this the probability of the optimal blocking. This information is unfortunately not available in the network.

To solve this problem, two significant challenges should be taken into account: (1) designing an access control strategy for dynamic generation of the blocking factor and (2) estimating the number of devices simultaneously attempting to access the network.

In this chapter, we tackle these questions using an estimator that was suggested in earlier research (Bouzouita *et al.* 2019). Since this estimation is very noisy, we exploit the potential of the most advanced reinforcement learning techniques, to take account of this complex reality (the state of the network is not observable) and deduce a sub-optimal control strategy. More especially in this chapter, we use the deep reinforcement learning algorithm Twin Delayed Deep Deterministic policy gradient algorithm (TD3) (Fujimoto *et al.* 2018) to produce the optimal blocking factor from these past estimations (Hadjadj-Aoul and Ait-Chellouche 2020).

The remainder of this chapter is organized as follows. Section 2.2 briefly presents the NB-IoT standard considered in our proposal. Section 2.3 gives an overview of the main congestion control techniques in IoT networks, and more especially cellular IoT networks. Section 2.4 describes the model for IoT terminals to access the network. Section 2.5 describes the suggested control solution, based on the TD3 algorithm, adapted to solve the problem of calculating the blocking factor. Section 2.6 describes the environment for simulating the suggested approach and shows its effectiveness compared to the existing approach. Finally, the chapter ends with a summary recapitulating the main advantages and achievements of the system suggested in section 2.7.

2.2. Fundamentals of the NB-IoT standard

NB-IoT is a cellular access technology of the Low Power Wide Area Network (LPWAN) type, specified in the document "Rel-13" by the 3GPP (3GPP 2016). Improvements to it were then provided in the documents "Rel-14", "Rel-15" and "Rel-16" (Mwakwata *et al.* 2019). Several key players in the industry, such as Ericsson, Nokia, Intel or Huawei, have shown a keen interest in this standard and have broadly participated in its standardization (Bicheno 2015).

NB-IoT technology is also presented as a promising candidate for covering one of the three main pillars of 5G, that is the massive Machine Type Communication (mMTC) (3GPP 2018a; Narayanan *et al.* 2018). In fact, the ITU has defined different requirements in terms of the density of the IoT terminals, battery life, coverage, price and mechanisms and features supported, to which the suggested technologies must respond to implement mMTCs (Kafle *et al.* 2016). In this direction, the 3GPP has suggested several mechanisms and NB-IoT features to meet these demands through different *releases* (Mwakwata *et al.* 2019). The options for integrating the NB-IoT with the core 5G network and its coexistence with the other services suggested in the framework of this standard are also studied by the 3GPP (3GPP 2019a). A better use of physical resources via the virtualization of network functions (NFV) or the SDN, as discussed in Migabo *et al.* (2020), can also facilitate this integration.

2.2.1. *Deployment and instances of use*

Many operators worldwide chose the NB-IoT standard from when it was standardized in 2016. Indeed, its properties, inherent to cellular technologies such as flexibility, adaptability, "over-the-air" service updates and cost containment, have accelerated adoption of this standard. According to a study by GSMA Intelligence, the total number of cellular IoT connections should reach 3.5 billion by 2025 (GSMA 2020). According to the same study, 94 commercial NB-IoT networks were listed in 2020. These latter are mainly located in China, where the standard is promised by the manufacturer Huawei. In Europe, Vodafone is the main operator promoting this standard in France, the operator SFR has integrated NB-IoT to its provision since last year.

This adoption of NB-IoT, by operators on the one hand and by manufacturers on the other, has occasioned many applications (Huawei 2017; Ray 2017), particularly in the Industry 4.0 domains, smart measurements, connected cities, monitoring and surveillance, smart agriculture and connected farms, or health.

2.2.2. *Transmission principles*

As its name indicates, NB-IoT is based on signal transmission over a narrow band. In fact, its spectral occupation is only 180 kHz, which is equivalent to an LTE Physical Resource Block (PRB). Because of this, it results in very low flow rates (at most 250 kbit/s), but these are sufficient for the applications for which this protocol is designed.

NB-IoT technology relies on LTE from which it inherits numerous features and mechanisms, especially at the level of physical and MAC layers. It also re-uses the same numerologies, channel coding, interleaving, etc. This has made it possible to reduce, on the one hand, the time needed to specify this standard and, on the other hand, the costs of developing the NB-IoT devices. However, since the bandwidth used is very narrow, modifications were needed to permit the primary objectives of this standard, that is, the massive use of long-range connections at reduced complexity and cost (Flore 2016).

2.2.2.1. *Modes of deployment*

As illustrated in Figure 2.2, NB-IoT can be deployed in three different modes: (1) in-band, within the LTE frequency band by substituting it for a PRB; (2) guard-band, by making use of unused spectrum resources, outside the traditional LTE frequency band; or finally (3) standalone, on an independent carrier. In this latter case, GSM frequencies are most often marked for deployment (Wang *et al.* 2017).

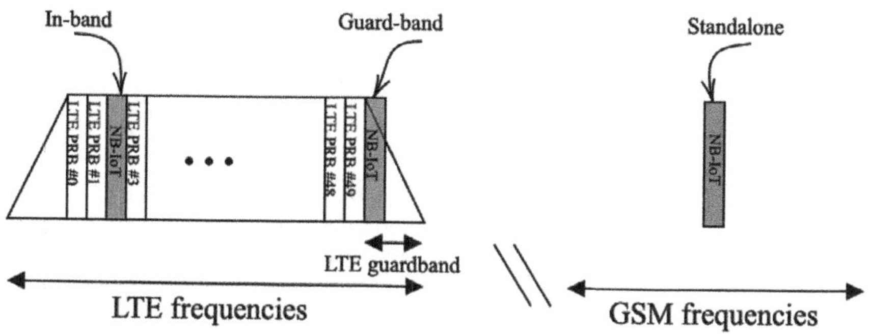

Figure 2.2. *NB-IoT deployment modes*

The three deployment modes cited above are transparent for non NB-IoT terminals. These consider the neighboring NB-IoT communications as noise.

2.2.2.2. Physical layer

On the downlink, transmissions are based on OFDM. Each OFDM symbol occupies 12 sub-carriers each of 15 kHz, thus occupying the equivalent of an LTE PRB (180 kHz). Also, in a similar way to LTE, each frame is composed of 10 sub-frames, each of two slots. The durations of the frame, the sub-frame and the slot are 10 ms, 2 ms and 0.5 ms, respectively. Each terminal operates on a Resource Block (RB) formed of seven consecutive OFDM symbols over the whole of the frequency domain. Concerning modulation, only Quadrature Phase-Shift Keying (QPSK) is supported.

On the uplink, transmissions are based on SC-FDMA (Single Carrier-Frequency Division Multiple Access). Two configuration modes are possible: single tone or multi-tone. When single tone is used, the two numerologies 15 kHz and 3.75 kHz are possible. The RU radio resource unit allocated to the terminals corresponds to the deviation from a sub-carrier. We thus have 12 or 48 RU, respectively. In multi-tone, only 15 kHz sub-carrier deviation can be used. The RU can occupy 3, 6 or 12 contiguous sub-carriers. The modulations used on the uplinks are BPSK and QPSK.

The signals and physical channels used in NB-IoT on the uplinks and the downlinks are summarized in Table 2.1 (Savaux *et al.* 2020).

	Physical channel/signal	Function
Downlink	*Narrowband Reference Signal* (NRS)	Dedicated to estimating the channel in the frequency domain (provides the reference stage). This signal is transmitted in all the sub-frames except for those reserved for NPSS and NSSS.
Downlink	*Narrowband Primary and Secondary Synchronization Signals* (NPSS and NSSS)	Inherited from LTE, these signals are used in discovering the cell via a time and frequency synchronization. They also make it possible to discover the identifier of the NCellID cell, notably via the NSSS.
Downlink	*Narrowband Physical Broadcast Channel* (NPBCH)	Transports the MIB-NB (Narrowband Master Information Block). This contains 34 data bits (information useful to the terminal, notably when decoding the SIB1-NB, the activation or not of the ACB, etc.) and 16 bits and CRC. The MIB-NB remains unchanged for a period 640 ms.
Downlink	*Narrowband Physical Downlink Control Channel* (NPDCCH)	Controls data transmissions between the base station and the NB-IoT terminal. It transports signaling information that governs these transmissions such as resource scheduling on the downlink, data acknowledgments sent on the uplink, the type of modulation used, etc.
Downlink	*Narrowband Physical Downlink Shared Channel* (NPDSCH)	Dedicated to the transmission of data in the direction of the terminal. It transports notably the SIB1-NB (Narrowband System Information Block) and some control information. SIB1-NB contains all the information useful for acquiring the other SIB blocks.
Uplink	*DeModulation Reference Signal* (DMRS)	Dedicated to the estimation of the channel in the frequency domain on the uplink. Unlike the NRS, this signal is always multiplexed with data. It is thus only transmitted in the RUs containing the data.
Uplink	*Narrowband Physical Random Access Channel* (NPRACH)	Dedicated to transmission of the preamble, which is the first signal transmitted by the terminal, to establish a connection to the network. Details of this channel are given in section 2.2.3.
Uplink	*Narrowband Physical Uplink Shared Channel* (NPDSCH)	Used for the transmission both of user data and of control data. The distinction is made using two formats: format 1 and format 2, respectively.

Table 2.1. *Physical channel and signals from the NB-IoT*

To permit an extension in coverage while still limiting the maximum transmission power of 23 dBm, NB-IoT devices use improved coverage techniques such as increasing power on the downlinks or the repetition of sub-frames both on the uplinks and on the downlinks. This technique is particularly helpful when the sensors are deployed in zones that are distant or difficult to access. The standard's target objective is to support a Maximum Coupling Loss (MCL) of at most 164 dB (20 dB more than LTE). The study of the extension of coverage envisaged by NB-IoT in different configurations, carried out in Adhikary *et al.* (2016), concludes with an MCL support above the 164 dB limit fixed by the 3GPP, that is, MCLs of 170.2 and 172.2 dB for respective spacings of 15 and 3.75 kHz, by carrying out 128 repetitions.

As in LTE, the network can define up to three levels of coverage extension (CE0, CE1 and CE2) to adapt to different radio conditions in which the terminals operate. These levels are defined across two power thresholds based on the power of the reference signal received by the NB-IoT (3GPP 2017a) terminals. Consequently, two thresholds can thus be defined per cell to configure the three levels of coverage. For each level, a number of repetitions of the signal are defined. The greater the power of the signal (less attenuation on the path), the lower the number of repetitions needed.

Unlike the LTE standard that supports the full duplex mode, the choice of half duplex type-B FDD was taken in Rel-13 for NB-IoT (3GPP 2017b). The terminals are therefore either transmitting, or receiving, but cannot carry out both operations simultaneously. In addition, a guard sub-frame is envisaged for each change from the uplink to the downlink or *vice versa*.

2.2.3. *Radio resource random access procedure*

The NPRACH channel has been completely rethought in NB-IoT not only to improve network coverage and energy consumption but also to adapt to the narrow bandwidth of NB-IoT (Lin *et al.* 2016). Indeed, in LTE the PRACH channel alone occupies more bandwidth than NB-IoT occupies in its entirety (1.08 MHz vs. 180 kHz).

Each NB-IoT terminal wishing to connect or resynchronize at the base station on the uplink after a long period of inactivity should execute a random access procedure. The first stage of this procedure consists of transmitting a preamble sequence on one of the frequency bands periodically allocated to the NPRACH channel. This is called Random Access Opportunity (RAO).

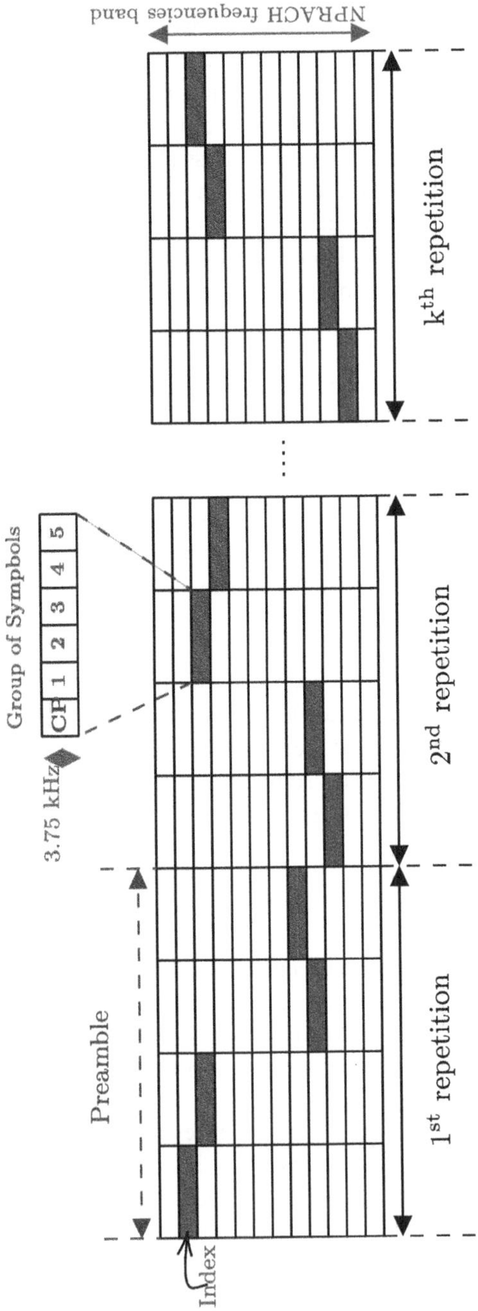

Figure 2.3. *Structure of a preamble sequence. For a color version of this figure, see www.iste.co.uk/chalouf/intelligent.zip*

The preamble consists of a set of four groups of OFDM symbols, as illustrated in Figure 2.3. This is the MCL, targeted by the standard, which determines the number 4. Each group of symbols is formed of a cyclical prefix (CP) and of K data symbols. To maintain the orthogonality of random access transmissions over different sub-carriers, the CP should be long enough to palliate long round trip times, especially in cells as large as those targeted in NB-IoT (up to 35 km) (3GPP 2015). The higher the number K, the more the overhead generated by the CP is reduced. On the contrary, K should be kept small so as to control interferences. In NB-IoT, K is fixed at 5 and two CP lengths are defined for both formats of the NPRACH channel, that is, 266.7 μs and 66.7 μs (Lin *et al.* 2016; 3GPP 2020).

For the purpose of extending coverage, the preamble can be repeated k times ($k = 2^i$, $i = 0, \ldots, 7$) (3GPP 2019b). The preamble sequence sent by the terminal is, consequently, formed of 4×2^i group symbols. The network can thus define up to three different configurations of the NPRACH resource per cell, depending on the classes of coverage considered. The number of repetitions k is thus defined for each of the configurations.

Each group of symbols is modulated over a different sub-carrier to the others. The NPRACH channel uses only single tone mode with a spacing of 3.75 kHz. A frequency band of up to 48 sub-carriers can thus be allocated to this channel with a base band of 12 sub-carriers. Thus, 12, 24, 36 or 48 contiguous sub-carriers are allocated to this channel in each coverage class. Consequently, the terminal has 12, 24, 36 or 48 orthogonal preambles and chooses one at random to transmit.

In addition, as illustrated in Figure 2.3, NB-IoT defines two jump models in the frequency band allocated to the NPRACH channel: (1) a fixed model for the jumps between the different groups of symbols forming the same preamble and (2) a pseudo-random model for repetitions of this preamble.

Thus, within a preamble, a frequency jump is applied between the first and the second group symbols and between the third and the fourth symbols. Another jump of six frequencies is also applied between the second and the third group of symbols. The pseudo-random model is, for its part, applied though by choosing start sub-carrier indexes from different repetitions of the preamble, by taking the cell identifier and the number of repetitions expected (Lin *et al.* 2016) as input data.

The random access procedure in NB-IoT, illustrated in Figure 2.4, is an exchange in four stages between the terminal and the base station (3GPP 2018b):

– the terminal transmits the preamble chosen at the first RAO opportunity and sets a timer for reception of the RAR response;

– if the preamble is clearly detected by the base station, the latter sends an RAR response conveying the timing advance and the allocated resource;

– the base station performs conflict resolution and sends the identity of the winning terminal in the conflict resolution message. If the message does not arrive at the terminal, the terminal continues to await it until the timer expires;

– the terminal then sends a connection request, by using the resource allocated to it and again sets a conflict resolution timer. This request, named msg3, conveys in particular the terminal identity.

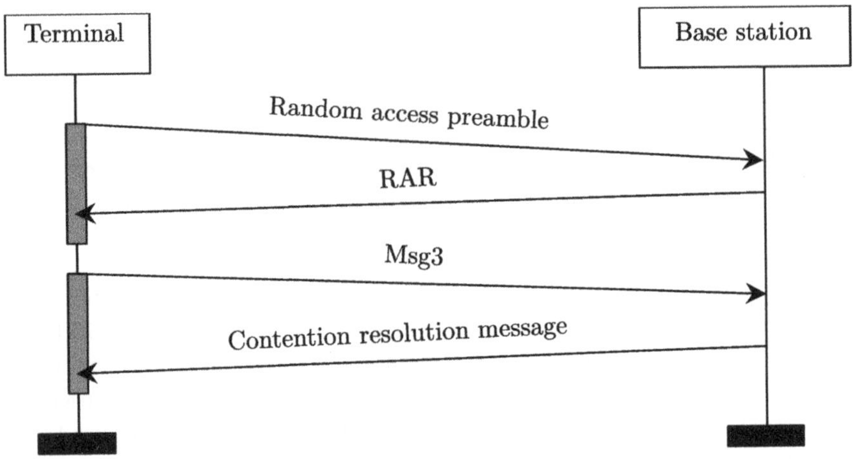

Figure 2.4. *Random access procedure*

This procedure fails if the terminal does not receive one of two responses from the base station in the windows of time defined by the two timers. Collisions between preambles, sent by different terminals, are often the cause of failure. In fact, if two terminals or more choose the same preamble on the same ROA opportunity, each will see its access attempt fail.

Each terminal for which the access procedure has failed observes a wait time chosen at random in a predefined interval and then retransmits its preamble. The number of retransmissions permitted depends on the terminal coverage class. If this number is reached and the terminal still does not succeed with the access procedure launched, this latter goes into the following coverage class if there is one or ends with a definitive failure of the access procedure.

2.3. State of the art

Beyond the generalization of broadband, 5G promises to improve our quality of life via connected ecosystems. The mMTC will play a determining role in carrying these out. Although this vision of the IoT seems very attractive, it also creates many challenges for network manufacturers and operators to tackle. In fact, an excessively high number of sensors will need to be operated in IoT objects in the coming years.

As explained in section 2.2.3, each terminal wishing to connect to the network should initiate a random access procedure. However, this procedure was initially designed for a limited number of terminals and the high density targeted by NB-IoT can lead very quickly to a situation of severe congestion. In fact, since the number of preambles available to each RAO is limited, the higher the number of terminals attempting access, the higher the risk of collision, thus leading to failure of the procedure for all the terminals that have chosen the same preamble. Certainly, the terminals that have not succeeded in access can retransmit the preamble after observing a wait time, but these retransmissions can also lead to a poor use of spectral resources, on the one hand, but also to increase energy consumption at the terminals on the other (Harwahyu *et al.* 2019).

Given its criticality, the random access procedure has been the subject of several studies. Some studies, such as Baracat and Brito (2018), Harwahyu *et al.* (2018) and Jiang *et al.* (2018), have suggested analytical models for optimizing the probability of success for access attempts at the terminals and the average access time in different configurations especially under time constraints (Harwahyu *et al.* 2018). Others are focused on retransmissions. So in Sun *et al.* (2017), a model based on Markov chains has been proposed to model the number of retransmissions; in Harwahyu *et al.* (2019), the authors propose a model to find a compromise between the number of repetitions predicted in the physical layer and the number of retransmissions expected in the MAC layer to optimize these two values by using the probability of preamble detection. The study has shown that the retransmissions considered in NPRACH can reduce the number of repetitions. These are only necessary when the network conditions deteriorate.

In Lin *et al.* (2016), Hwang *et al.* (2018) and Jeon *et al.* (2018), the focus has been on transmission of the preamble and estimation of the arrival time. Thus, a detection algorithm on the receiver side, a new jump model in the NPRACH frequency domain and a framework to detect multiple users have been proposed, respectively. In Zhang *et al.* (2020), the TA preambles that have undergone a collision are used to improve performances of the random access procedure.

From the perspective of standardization, congestion control at the level of access to the network was very early on identified as a priority by the 3GPP and ETSI organisms (3GPP 2011). The cellular IoT, especially NB-IoT, therefore benefits naturally from the solutions proposed for the standards that precede them. Among the solutions proposed, we find ACB (Access Class Barring) and EAB extension, slotted random access, backoffs specific to MTC, a dynamic allocation of resources, etc. (Ali *et al.* 2017).

The ACB and EAB are those that tackle the problem at its root by blocking access to the network via the diffusion of blocking parameters in the system's information blocks (SIB) blocks at each RAO. The terminals receive, especially, a probability of blocking p and a blocking time T_b for this opportunity. Each terminal wanting to access the network generates an access probability q. If $q < p$, the terminal has permission to make an access attempt, otherwise this is postponed for a time T_b. This mechanism has been extended. In the EAB, the terminals are classed according to their requirements in terms of Quality of Service (QoS) and the EAB algorithm dynamically blocks low-priority terminals according to the arrival rate by diffusing a *bitmap* in the SIB14.

It seems clear that congestion control via these techniques relies entirely on the probability of blocking defined by the network. In fact, if the probability of blocking is too high, then a significant number of terminals would pass the access control, thus leading to collisions, and if, on the other hand, this probability is too small, then collisions will be reduced, but a large number of terminals will switch to the inactive mode and this will lead to under-use of resources. It is, therefore, essential to calculate a probability of optimal blocking for effective congestion management.

A study of ACB and EAB performances was carried out in Toor and Jin (2017). Comparison of the two techniques via simulation has shown that the ACB is best adapted to communications with high delay constraints and the EAB in the case of energy-constrained terminals. However, calculating the probability of optimal blocking relies on the base station ability to know the number of terminals attempting the access to the network. In practice, this is not the case. In fact, the base station does not have knowledge of the number of terminals whose access attempts have been blocked.

Several mechanisms have been proposed to estimate the number of terminals attempting to access the network (including the terminals blocked by the access control) so as to deduce from this the blocking probability to be used. In Park and Lim (2016), in the absence of knowledge of the number of blocked terminals, the authors use a heuristic to adapt the probability of blocking. The algorithm proposed in Liu *et al.* (2020) makes a recursive Bayesian estimation of the active terminals in each class, and depending on this estimation some preambles are allocated to different classes. The algorithm was then improved by assigning an ACB blocking factors to each of the classes, independently of the others, for better congestion control. In Jin *et al.* (2017), a recursive Bayesian estimation of the active terminals, based on the number of unchosen preambles, makes it possible to calculate a blocking factor for those arrivals from the terminals that are sporadic in character. Performance of the EAB technique is studied in Cheng *et al.* (2015) for LTE-A networks. The optimal values of the radio messaging cycle as well as the periodicity of SIB14 are then derived by submitting the analytic model to targeted QoS constraints.

In this chapter, we base ourselves on an estimator proposed in an earlier work (Bouzouita *et al.* 2019) and unlike the work cited above, we use reinforcement learning techniques, especially the TD3 algorithm, to calculate an optimal blocking factor from a set of past estimations. To our knowledge, this is the first time that this type of algorithm has been used in managing massive access to terminals in NB-IoT networks, excepting our previous contribution (Hadjadj-Aoul and Ait-Chellouche 2020).

2.4. Model for accessing IoT terminals

The model proposed represents an overview of IoT devices executing the ACB algorithm. During the random access attempt, the IoT devices compete for the same available preambles. As the 3GPP standard indicates, the number of preambles N should be an integer as explained in section 2.3 (ETSI 2011).

During each access opportunity (i.e. Random Access Channel, RACH), these preambles are divided into successful preambles, chosen by a single device, in collision, chosen by two or more devices, and free, not chosen by any of these devices.

In what follows, we calculate the average values of these quantities, which we determined in Bouzouita *et al.* (2015). These will thereafter be used by our algorithms.

Let us define $q_N = 1 - 1/N$. The average number of successful preambles N_S, during RACH opportunities, is given as follows (this is a classic problem of throwing balls into bins):

$$N_S = q_N^{x_2-1} x_2, \qquad [2.1]$$

where x_2 represents the number of devices attempting access. As we have been able to demonstrate in Bouzouita *et al.* (2019), equation [2.1] is maximized (i.e. derivative equals zero) when the number of devices x_2^* attempting access simultaneously is

$$x_2^* = -\frac{1}{\ln q_N}. \quad [2.2]$$

The average number of free preambles N_I is given by equation [2.3]:

$$N_I = Nq_N^{x_2} \quad [2.3]$$

From [2.1] and [2.2], we obtain the expected number of failed preambles N_F:

$$N_F = N - (N_S + N_I) \quad [2.4]$$

The modeled system is an approximation of reality in many ways, especially where it concerns the fixed and limited number of access attempts. However, we preferred to simplify the model to make it more tractable (Figure 2.5).

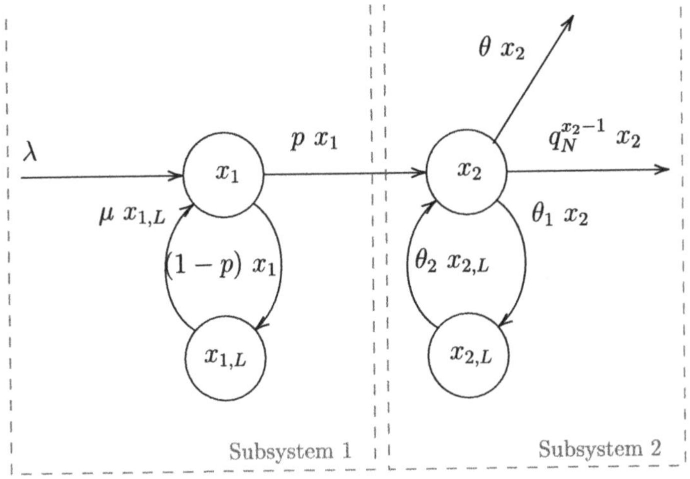

Figure 2.5. *System model. For a color version of this figure, see www.iste.co.uk/chalouf/intelligent.zip*

The sub-system 1 represents the terminals that wish to connect; the IoT devices belonging to the state variable x_1 represent those that can try to connect with a probability p; in case of failure, they go into the waiting state $x_{1,L}$ for a set duration. Sub-system 2 represents the IoT devices that can try to choose a preamble. In case of collision, they can try to access it several times. Each quits sub-system 2 when it succeeds in being the only one to have chosen a preamble or when it reaches the maximum number of attempts (with a rate of θ).

The proposed model is fluid: the quantities involved and the whole numbers are considered to be real (continuous) quantities. The parameters used are listed below:

– $x_1(t)$ is the number of waiting devices;

– $x_{1,L}(t)$ is the number of blocked devices, after a failure at the end of an access attempt (i.e. ACB);

– $x_2(t)$ is the total number of devices that pass the ACB control and wait to start the random access attempt (RA);

– $x_{2,L}(t)$ is the number of blocked devices, at instant t, after a failed RA attempt, awaiting a new attempt;

– λ is the arrival rate of IoT objects;

– μ is the rate of objects that can reattempt ACB after a failure;

– θ_1 is the RA failure rate, which is equal to $1 - q_N^{x_2-1}$ when θ is equal to 0 (see the point before last);

– θ_2 is the rate of objects that can attempt access after a failure;

– θ is the rate at which the device abandons transmission after having reached the maximum number of RA attempts; in a correctly dimensioned system, we should have $\theta = 0$;

– p is the ACB blocking factor.

We are now able to describe the evolution of state variables $x_1(t)$, $x_{1,L}(t)$, $x_2(t)$ and $x_{2,L}(t)$ basing our study on the model represented in Figure 2.5. The model's dynamic is described by the following system of differential equations:

$$\begin{cases} \dfrac{dx_1}{dt} &= \lambda - x_1 + \mu x_{1,L} \\ \dfrac{dx_{1,L}}{dt} &= (1-p)x_1 - \mu x_{1,L} \\ \dfrac{dx_2}{dt} &= px_1 - \left(\theta + \theta_1 + q_N^{x_2-1}\right)x_2 + \theta_2 x_{2,L} \\ \dfrac{dx_{2,L}}{dt} &= \theta_1 x_2 - \theta_2 x_{2,L} \end{cases} \qquad [2.5]$$

In what follows, we suppose that $\theta = 0$, to simplify the model. In fact, a system where the devices often reach the maximum number of attempts is an unstable system, which we naturally try to avoid.

The model described in [2.5] is nonlinear and not affine, in the control. It can easily be demonstrated that the model described is not observable, in view of its state $[x_1 \quad x_{1,L} \quad x_2 \quad x_{2,L}]$ which cannot be known precisely. It is also not controllable, as the blocking factor p can only act partially on the state. These properties render the synthesis of an optimal controller, guaranteeing the stability of the very complex system, described above, very complex.

Although the state is not observable, it is possible to produce an estimation of the average number of devices attempting access $\hat{x}_2$ by reversing equations [2.1] and [2.3]. This gives a very noisy measurement, but one which is, nevertheless, useful for IoT blocking as we demonstrated in Bouzouita *et al.* (2019).

2.5. Access controller for IoT terminals based on reinforcement learning

The difficulty of observing the system state, described in section 2.4, has led us to consider strategies making it possible to deduce the blocking factor even in the presence of very noisy measurements.

It is in this sense that we relied on deep learning techniques, which demonstrated great effectiveness in automatically extracting characteristics of system "features" in the presence of data tainted with noise or even of incomplete data (Rolnick *et al.* 2017).

Given the lack of data, we have considered the class of reinforcement learning techniques.

More particularly, we considered the "Twin Delayed Deep Deterministic policy gradient algorithm" (TD3) technique, which can tackle a continuous action space, and which has shown greater effectiveness in learning speed and in performance than existing approaches (Fujimoto *et al.* 2018).

We formulate, in what follows, the problem of access in the IoT as a reinforcement learning problem, in which an agent finds iteratively a sub-optimal blocking factor, making it possible to reduce the access conflict.

2.5.1. *Formulating the problem*

In reinforcement learning (Sutton *et al.* 2019), we have two main entities, an environment and an agent. The learning process happens through interaction between these entities so that the agent can optimize a total revenue. At each stage t, the agent obtains a representation of the state s_t of the environment and chooses an action a_t, based on this. Then the agent applies this action to the environment. Consequently, the environment passes to a new state s_{t+1} and the agent receives a recompense r_t. This interaction can be modeled as a Markovian decision process (MDP) $M = (S, A, P, R)$, where S is the state space, A the action space, P the dynamic transition and R the revenue function. The behavior of the agent is defined by its policy $\pi: S \longrightarrow A$, which makes it possible to link a state to an action where a deterministic system is involved, or a distribution of action where it is probabilistic. The objective of such a system is to find the optimal policy $\pi^\star$, making it possible to maximize the accumulated revenue.

In the problem of controlling access to the IoT, we define a discrete MDP, where the state, the action and the revenue are defined as follows:

– The state: given that the number of terminals attempting access at a given instant k is unavailable, the state we are considering is based on measured estimates. Since a single measurement of this number is necessarily very noisy, we will consider a set of several measurements, which can better reveal the state present in the network. The state s_k is defined as the vector $(\hat{x}_2^k, \hat{x}_2^{k-1}, \cdots, \hat{x}_2^{k-H})$, where H represents the measurement horizon.

– The action: at each state, the agent must select the blocking factor p which will need to be considered by the IoT objects. This value is continuous and determinist in the problem that we consider, that is that the same state s_k will always give the same action a_k.

– The revenue: this is a signal that the agent receives from the environment after the execution of an action. Thus, at stage k, the agent obtains a revenue r_k as a consequence of the action a_k that it carried out in state s_k. This revenue will allow the agent to know the quality of the action executed, the objective of the agent being to maximize this revenue.

Unlike classical reinforcement learning problems, the optimum is known here and given by equation [2.2]. The revenue is given by equation [2.6]:

$$r_k = e^{-|x_2^* - \hat{x}_2^k|} \qquad [2.6]$$

The revenue is therefore maximal when the chosen action makes it possible to obtain a number of devices attempting access ($\hat{x}_2^k$) equal to the optimum x_2^*. However, as the measurement $\hat{x}_2^k$ is marred by noise, this impacts the measured revenue.

The objective of such a system is to find the blocking probability, making it possible to maximize the average recompense, which amounts to reducing the distance between the measurements of the number of terminals attempting access and the optimum. To meet this objective, we rely on the TD3 algorithm.

The TD3 algorithm is an actor-critic approach, where the actor is a network of neurons which decides the action to take in a particular state; the main network makes it possible to know the value of being in a state and to choose a particular action. TD3 makes it possible to resolve the question of over-evaluation in estimating the value (Thrun and Schwartz 1993) by introducing two critical networks and by taking the minimum between these two estimations. This approach is particularly beneficial in our case due to the inherent presence of measurement errors.

2.5.2. *Regulation system for arrivals*

The diagram in Figure 2.6 describes the system that makes it possible to control the number of attempts from IoT objects. This system is based on the diffusion of the blocking factor at the terminals, through the SIBs which are broadcasted, and more particularly through the Type14 SIB block, which makes it possible to diffuse the access blocking parameters (ETSI 2019).

Following the reception of the blocking factor, the terminals wishing to carry out transmission execute the ACB, which allows them to pass to the following stages with a probability p, which is calculated by our TD3 based controller. These terminals can, consequently, attempt access by choosing a preamble at random from among the available preambles. Knowing the state of the preambles, the gNodeB can estimate the number of attempts made. This measure is very noisy, since the model given only makes it possible to estimate averages. We take an average estimate of the number of devices. We use a sliding average to do this.

The controller, we have proposed, receives these measurements, augmented from the revenue, at the end of each preamble. The revenue obtained enables it to know the quality of the actions taken. These different data are placed in a memory of past experiences. This is a random sub-set of this memory that will enable it to learn robustly and to choose, subsequently, a new action.

These different actions are repeated cyclically.

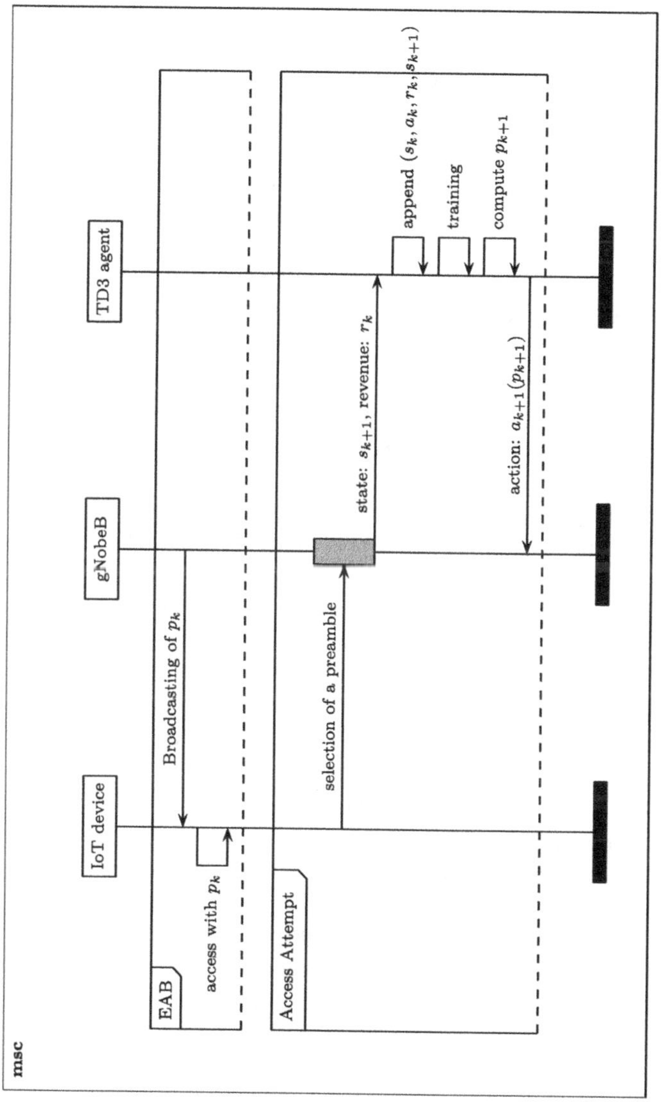

Figure 2.6. *System for regulating arrivals*

2.6. Performance evaluation

After having described our access controller proposition, in this section we evaluate its performances, using a simulation environment that we have built in Simpy (2020).

We have considered an NB-IoT antenna in which access requests arrive according to a Poisson distribution with an average rate between two arrivals of 0.018 s. We have considered a number of preambles N equal to 16, with an arrival frequency equal to 0.1 s. In the system considered, each device attempting access will be able to do so a maximum of 16 times. Beyond this limit, the terminal abandons transmission.

Our controller's performance, which is based on the TD3 technique, is compared to an adaptive approach. We have considered a measurement horizon H equal to 10. Use of a larger measurement window does not allow a significant improvement in performances, which means that a window of 10 measurements makes it possible to reflect sufficiently the real state of the network.

The adaptive approach consists of gradually increasing the blocking probability when the number of attempts is beyond a predefined threshold above the optimal value. When a value is below a predefined threshold below the optimal value, the blocking probability is gradually reduced, to allow more terminals to attempt access.

In Figures 2.7 and 2.8, the blocking probabilities for both strategies considered are expressed. The adaptive technique (Figure 2.7) starts with an access probability of 1 and adapts itself according to the traffic conditions, which change following a Poisson distribution. For the strategy, which is based on the TD3 algorithm, there is an initial stage lasting 200 s, where the algorithm tries to explore the action space according to a uniform law (Figure 2.8). It is only after this stage that the algorithm begins to make use of its learning, which is refined in line with its experiences.

We can note that under TD3 (Figure 2.8), future actions have no links with past actions, unlike the adaptive case. In fact, the values of the actions can change completely, because they depend only on the state of the network, which can change very quickly.

Figures 2.9 and 2.10 describe the impact of control laws, described previously, on the average latency of the access attempts. In these plots, we do not consider the terminals that have abandoned transmission of sets of a number of maximum attempts. Even though we can note, in Figure 2.10, some terminals with latencies slightly higher than those in Figure 2.9, the latency is globally of the same order, that is, the TD3 algorithm does not show any advantage in terms of latency.

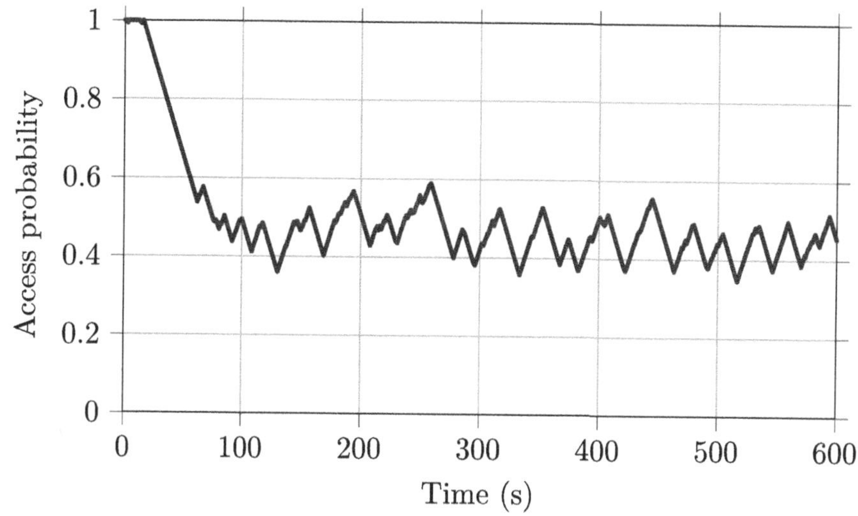

Figure 2.7. *Access probability with the adaptive controller*

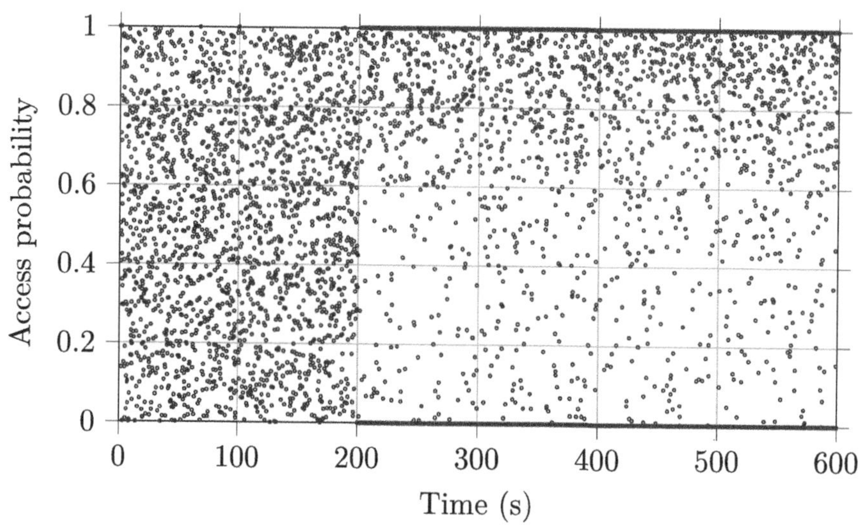

Figure 2.8. *Access probability with the controller using TD3*

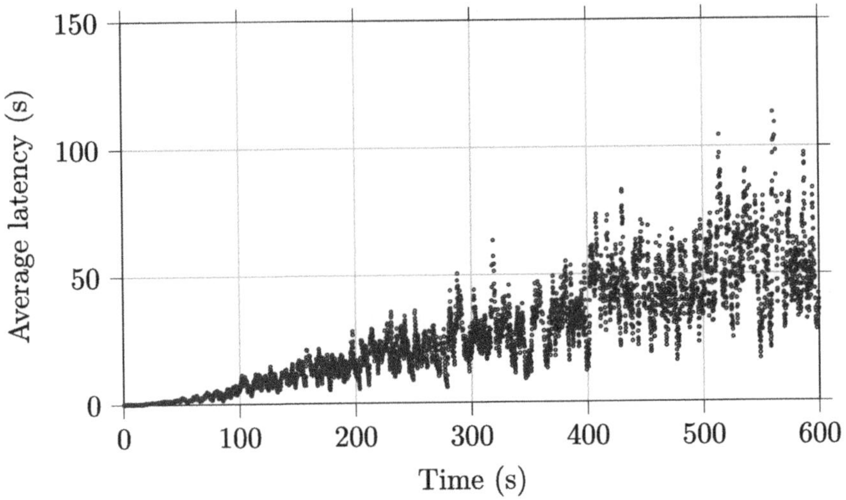

Figure 2.9. *Average latency of the terminals with the adaptive controller*

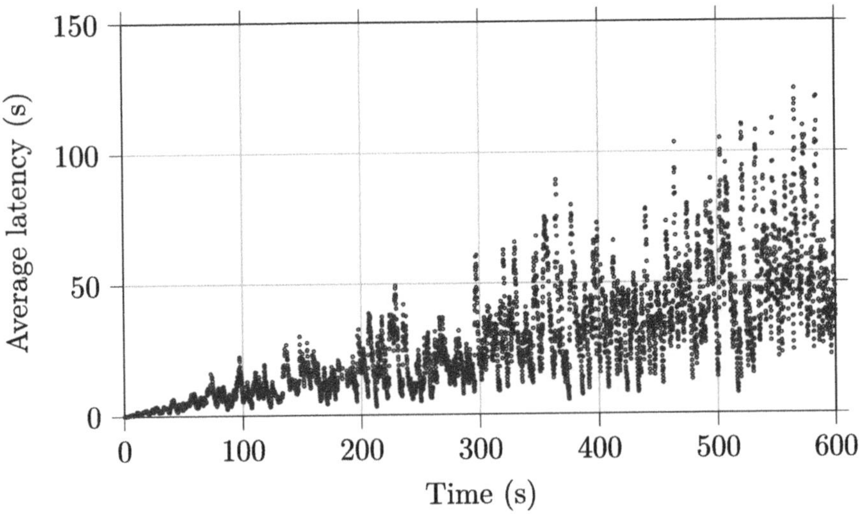

Figure 2.10. *Average latency of the terminals with the controller using TD3*

Even though TD3 does not show any particular advantage in terms of latency, we can see in Figure 2.12 that after an exploration stage, the revenue improves very significantly. This recompense is clearly higher than for the adaptive controller, which shows a reduced and very variable recompense (Figure 2.11). In fact, the average of the recompense in TD3 is in the order of 13.91%, while the adaptive controller shows a recompense in the order of 3.6%. This recompense reflects the fact that under TD3, the average number of terminals attempting access gets closer to the optimum. This result, perhaps also shown in Figure 2.14, shows that the number of attempts with TD3 is closer still to the optimum which is equal to 15.49. In fact, the average number of attempts using the adaptive controller is equal to 30.12 (Figure 2.13), while it is equal to 19.6 for our approach.

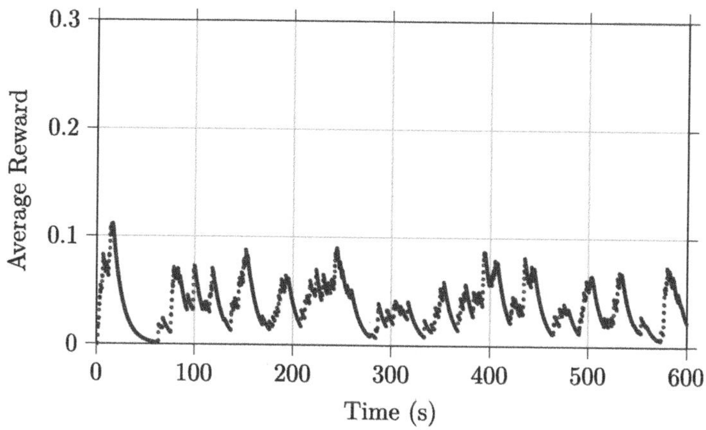

Figure 2.11. *The average recompense with the adaptive controller*

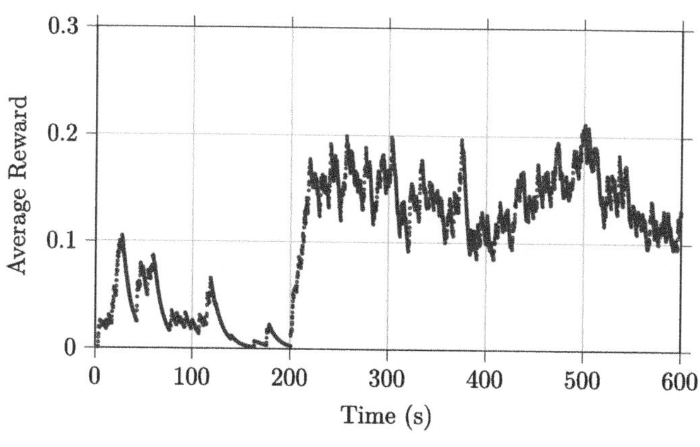

Figure 2.12. *The average recompense with the controller using TD3*

We can note that in Figure 2.13 the adaptive technique does not make it possible to control correctly the number of attempts. In fact, we very often reach numbers significantly higher than the optimum. This triggers many collisions at access and new access attempts. We also see that the number of abandonments remains relatively significant compared to the TD3 controller (Figure 2.14). The latter, after the exploration stage, succeeds in significantly reducing the number of abandonments, which demonstrates the effectiveness of the proposed approach.

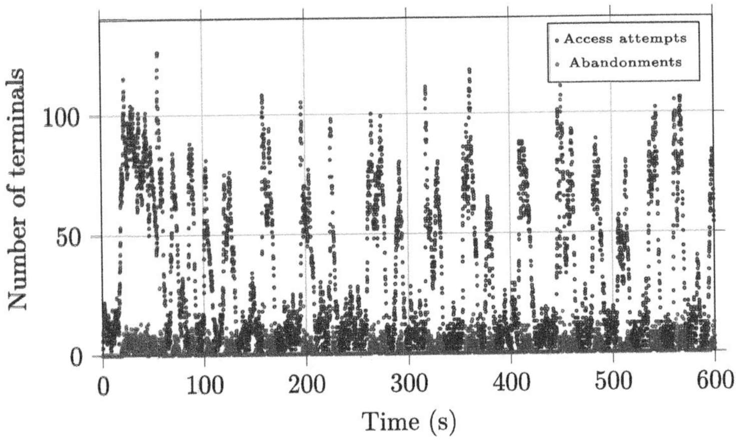

Figure 2.13. *Access attempts (blue) and abandonments (red) with the adaptive controller. For a color version of this figure, see www.iste.co.uk/chalouf/intelligent.zip*

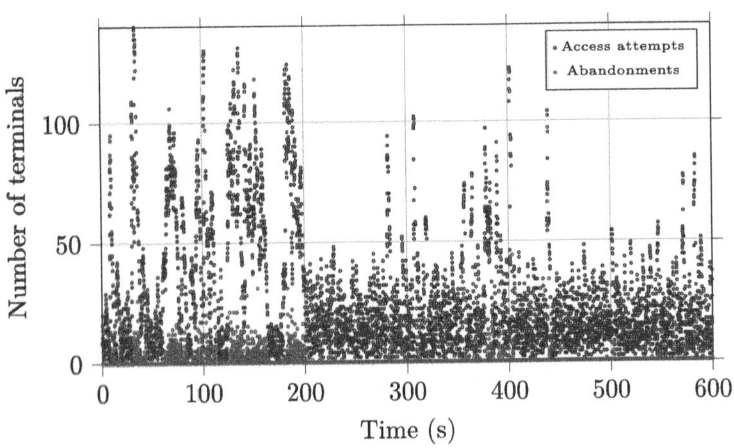

Figure 2.14. *Access attempts (blue) and abandonments (red) with the controller using TD3. For a color version of this figure, see www.iste.co.uk/chalouf/intelligent.zip*

It should be noted that having recourse to our approach based on reinforcement learning, we have an improvement in performance with each access attempt. The limit, however, is found in the estimation errors, which lead to errors in calculating the recompense, hence the importance of having precise estimators.

2.7. Conclusion

In this chapter, we proposed a mechanism to control congestion of the access network, which is considered one of the most critical problems for IoT objects. We have proposed tackling congestion at its root by effectively managing random accesses from these devices thanks to use of the ACB mechanism.

The proposed access control mechanism is different from conventional methods, which generally rely on simple heuristics. Indeed, the proposed technique relies on recent advances in deep reinforcement learning, through use of the TD3 algorithm. The proposed approach has, in addition, the advantage of learning from its environment and could therefore enable it to adapt to variation of the access schema.

The simulation results make it possible to show the superiority of the proposed approach, which succeeds in maintaining a number of access attempts close to the optimum, despite the absence of exact information on the number of access attempts. This work also makes it possible to show the potential of using learning techniques in environments where the state cannot be known with precision.

In the context of our future work, we envisage improving estimation of the number of attempts using learning techniques.

2.8. References

3GPP (2011). RAN improvements for machine-type communications. TR 37.868, version 11.0.0 [Online]. Available at: https://portal.3gpp.org/desktopmodules/Specifications/SpecificationDetails.aspx?specificationId=2630 [Accessed 14 October 2020].

3GPP (2015). Cellular system support for ultra-low complexity and low throughput Internet of Things (CIoT). TR 45.820 [Online]. Available at: https://portal.3gpp.org/ChangeRequests.aspx?q=1&versionId=46751&release=187 [Accessed 14 October 2020].

3GPP (2016). E-UTRA and E-UTRAN; LTE physical layer, general description (Release 13). TR TS 36.201 [Online]. Available at: https://portal.3gpp.org/desktopmodules/Specifications/SpecificationDetails.aspx?specificationId=2424 [Accessed 14 October 2020].

3GPP (2017a). Physical layer measurements (Release 13). TS 36.214, version 13.5.0 [Online]. Available at: https://portal.3gpp.org/desktopmodules/Specifications/SpecificationDetails.aspx?specificationId=2428 [Accessed 14 October 2020].

3GPP (2017b). User equipment (UE) radio transmission and reception (Release 13). TS 36.101, version 13.9.0 [Online]. Available at: https://portal.3gpp.org/desktopmodules/Specifications/SpecificationDetails.aspx?specificationId=2411 [Accessed 14 October 2020].

3GPP (2018a). Consideration on self-evaluation of IMT-2020 for mMTC connection density. R1-1801796 [Online]. Available at: http://www.3gpp.org/ftp/TSG_RAN/WG1_RL1/TSGR1_92/Docs/R1-1801796.zip [Accessed 14 October 2020].

3GPP (2018b). Radio Resource Control (RRC); protocol specification. TS 36.331 [Online]. Available at: https://portal.3gpp.org/desktopmodules/Specifications/SpecificationDetails.aspx?specificationId=2440 [Accessed 14 October 2020].

3GPP (2019a). Study on cellular Internet of Things (IoT) support and evolution for the 5G System. TR 23.724, version 16.1.0 [Online]. Available at: http://www.3gpp.org/DynaReport/23724.htm [Accessed 14 October 2020].

3GPP (2019b). Medium Access Control (MAC) protocol specification. TS 36.321, V14.12.0 [Online]. Available at: https://portal.3gpp.org/desktopmodules/Specifications/SpecificationDetails.aspx?specificationId=2437 [Accessed 14 October 2020].

3GPP (2020). Evolved Universal Terrestrial Radio Access (E-UTRA); physical channels and modulation (Release 16). TS 36.211, version 16.2.0, June 2020 [Online]. Available at: https://portal.3gpp.org/desktopmodules/Specifications/SpecificationDetails.aspx?specificationId=2425 [Accessed 14 October 2020].

Adhikary, A., Lin, X., Wang, Y.-P.-E. (2016). Performance evaluation of NB-IoT coverage. In *The IEEE 84th Vehicular Technology Conference (VTC-Fall)*. IEEE, Montreal.

Ali, M.S., Hossain, E., Kim, D.I. (2017). LTE/LTE-A random access for massive machine-type communications in smart cities. *IEEE Communications Magazine*, 55(1), 76–83.

Baracat, G. and Brito, J. (2018). NB-IoT random access procedure analysis. In *The Latin-American Conference on Communications (LATINCOM)*. IEEE, Guadalajara.

Bicheno, S. (2015). Intel, Nokia and Ericsson collaborate on NB-LTE wireless for IoT [Online]. Available at: https://telecoms.com/441951/intel-nokia-and-ericsson-collaborate-on-nb-lte-wireless-for-iot/ [Accessed 1 February 2022]

Bouzouita, M., Hadjadj-Aoul, Y., Zangar, N., Rubino, G., Tabbane, S. (2015). Multiple access class barring factors algorithm for M2M communications in LTE-advanced networks. In *18th ACM International Conference on Modeling, Analysis and Simulation of Wireless and Mobile Systems, MSWiM'15*. ACM, New York.

Bouzouita, M., Hadjadj-Aoult, Y., Zangar, N., Tabbane, S. (2016). On the risk of congestion collapse in heavily congested M2M networks. In *2016 International Symposium on Networks, Computers and Communications (ISNCC)*. IEEE, Yasmine Hammamet.

Bouzouita, M., Hadjadj-Aoul, Y., Zangar, N., Rubino, G. (2019). Estimating the number of contending IoT devices in 5G networks: Revealing the invisible. *Transactions on Emerging Telecommunications Technologies*, 30(4), e3513.

Cheng, R.C., Chen, J., Chen, D.W., Wei, C.H. (2015). Modeling and analysis of an extended access barring algorithm for machine-type communications in LTE-A Networks. *IEEE Transactions on Wireless Communications*, 14(6), 2956–2968.

ETSI (2011). LTE; Evolved Universal Terrestrial Radio Access (E-UTRA); physical channels and modulation. TR ETSI TS 136.211, version 10.0.0 [Online]. Available at: https://www.etsi.org/deliver/etsi_TS/136200_136299/136211/15.02.00_60/ts_136211v150200p.pdf [Accessed 14 October 2020].

ETSI (2019). LTE; Evolved Universal Terrestrial Radio Access (E-UTRA); Radio Resource Control (RRC); protocol specification (3GPP TS 36.331, version 14.12.0 Release 14). TS 136 331, version 14.12.0 [Online]. Available at: https://www.etsi.org/deliver/etsi_ts/136300_136399/136331/13.00.00_60/ts_136331v130000p.pdf [Accessed 14 October 2020].

Flore, D. (2016). 3GPP standards for the Internet-of-Things. *gSMA MIoT* [Online]. Available at: http://www.3gpp.org/news-events/3gpp-news/1766-iot_progress [Accessed 14 October 2020].

Fujimoto, S., van Hoof, H., Meger, D. (2018). Addressing function approximation error in actor-critic methods. In *The 35th International Conference on Machine Learning*. ICML, Stockholm.

GSMA Intelligence (2020). Mobile IoT (LPWA) [Online]. Available at: https://www.gsma.com/iot/mobile-iot/ [Accessed 14 October 2020].

Hadjadj-Aoul, Y. and Ait-Chellouche, S. (2020). Access control in NB-IoT networks: A deep reinforcement learning strategy. *Information* 11(11), 541 [Online]. Available at: https://doi.org/10.3390/info11110541.

Harwahyu, R., Cheng, R.G., Wei, C.H., Sari, R.F (2018). Optimization of random access channel in NB-IoT. *IEEE Internet of Things Journal*, 5(1), 391–402.

Harwahyu, R., Cheng, R.G., Tsai, W.J., Hwang, J.K., Bianchi, G. (2019). Repetitions vs Retransmissions: Trade-off in configuring NB-IoT random access channels. *IEEE Internet of Things Journal*, 6(2), 3796–3805.

Huawei (2017). NB-IoT commercial premier use case library [Online]. Available at: https://www.gsma.com/iot/wp-content/uploads/2017/12/NB-IoT-Commercial-Premier-Use-case-Library-1.0_Layout_171110.pdf [Accessed 14 October 2020].

Hwang, J.K., Li, C.F., Ma, C. (2018). Efficient detection and synchronization of superimposed NB-IoT NPRACH preambles. *IEEE Internet of Things*, 6, 1173–1182.

Jeon, W.S., Seo, S.B., Jeong, D.G. (2018). Effective frequency hopping pattern for ToA estimation in NB-IoT random access. *IEEE Transactions on Vehicular Technology*, 67(10), 10150–10154.

Jiang, N., Deng, Y., Condoluci, M., Guo, W., Nallanathan, A., Dohler, M. (2018). RACH preamble repetition in NB-IoT network. *IEEE Communications Letters*, 22(6), 1244–1247.

Jin, H., Toor, W.T., Jung, B.C., Seo, J.B. (2017). Recursive pseudo-Bayesian access class barring for M2M communications in LTE systems. *IEEE Transactions on Vehicular Technology*, 66(9), 8595–8599.

Kafle, V.P., Fukushima, Y., Harai, H. (2016). Internet of Things standardization in ITU and prospective networking technologies. *IEEE Communications Magazine*, 54(9), 43–49.

Lin, X., Adhikary, A., Wang, Y.-E. (2016). Random access preamble design and detection for 3GPP narrowband IoT systems. *IEEE Wireless Communications Letters*, 5(6), 640–643.

Liu, J., Agiwal, M., Qu, M., Jin, H. (2020). Online control of preamble groups with priority in massive IoT networks. *IEEE Journal on Selected Areas in Communications*. doi:10.1109/JSAC.2020.3018964.

Migabo, E.M., Djouani, K.D., Kurien, A.M. (2020). The narrowband Internet of Things (NB-IoT) resources management performance state of art, challenges, and opportunities. *IEEE Access*, 8, 97658–97675, doi: 10.1109/ACCESS.2020.2995938.

Mwakwata, C.B., Malik, H., Mahtab Alam, M., Le Moullec, Y., Parand, S., Mumtaz, S. (2019). Narrowband Internet of Thing (NB-IoT): From Physical (PHY) and Media Access Control (MAC) layers perspectives. *Sensors*, 19(11), 2613.

Narayanan, S., Tsolkas, D., Passas, N., Merakos, L. (2018). NB-IoT: A candidate technology for massive IoT in the 5G era. In *IEEE 23rd International Workshop on Computer Aided Modeling Design of Communication Links and Network (CAMAD)*. IEEE, Barcelona.

Park, A. and Lim, X. (2016). Adaptive access class barring method for machine generated communications. *Mobile Information Systems*, 1–6. doi:10.1155/2016/6923542.

Ray, B. (2017). NB-IoT case studies. Link-labs [Online]. Available at: https://www.link-labs.com/blog/nb-iot-case-studies [Accessed 14 October 2020].

Rolnick, D., Veit, A., Belongie, S.J., Shavit, N. (2017). Deep learning is robust to massive label noise [Online]. Available at: https://arxiv.org/abs/1705.10694 [Accessed 14 October 2020].

Savaux, V., Le Guen, M., Kanj, M. (2020). A tutorial on NB-IoT physical layer design. *IEEE Communications Surveys & Tutorials*, 22(4).

Simpy (2020). Discrete event simulation for Python [Online]. Available at: https://simpy.readthedocs.io/en/latest/ [Accessed 14 October 2020].

Sun, Y., Tong, F., Zhang, Z., He, S. (2017). Throughput modeling and analysis of random access in narrow-band Internet of Things. *IEEE Internet of Things*, 5(3), 1485–1493.

Sutton, R.S. and Barto, A.G. (2019). *Reinforcement Learning: An Introduction*, 2nd edition. MIT Press, Cambridge.

Thrun, S. and Schwartz, A. (1993). Issues in using function approximation for reinforcement learning. In *Proceedings of the 1993 Connectionist Models Summer School*, Mozer, M., Smolensky, P., Touretzky, D., Elman, J., Weigend, A. (eds). Lawrence Erlbaum, Hillsdale.

Toor, W.T. and Jin, H. (2017). Comparative study of access class barring and extended access barring for machine type communications. In *International Conference on Information and Communication Technology Convergence (ICTC)*. KICS, IEEE, Jeju Island, 12.

Wang, Y.-P.-E., Lin, X., Adhikary, A., Grovlen, A., Sui, Y., Blankenship, Y., Bergman, J., Razaghi, H.S. (2017). A primer on 3GPP narrowband Internet of Things. *IEEE Communications Magazine*, 55(3), 117–123.

Zhang, J., Xie, D., Wang, X. (2020). TARA: An efficient random access mechanism for NB-IoT by exploiting TA value difference in collided preambles. *IEEE Transactions on Mobile Computing*. doi: 10.1109/TMC.2020.3019224.

3

Optimizing Performances in the IoT: An Approach Based on Intelligent Radio

Badr BENMAMMAR
LTT, University Abou Bekr Belkaid Tlemcen, Algeria

3.1. Introduction

The Internet of Things (IoT) is a term used to mean the omnipresent, intelligent and interconnected objects that interact with one another with the help of standard communication protocols. The IoT signifies the omnipresence of various objects around us, such as sensors, Radio Frequency Identification (RFID), mobile telephones and others. The IoT will have a marked impact on daily life in the future, such as assisted living, automation, industrial manufacturing, logistics, process management, intelligent transport of people or materials, real-time monitoring of industrial processes and e-health facilities (Aijaz and Aghvami 2015).

IoT objects will be interconnected via wired and wireless communication technologies. However, accessibility problems for distant users make wireless communication the most attractive solution. Because the wireless spectrum is a precious resource, IoT applications can suffer from spectrum scarcity.

To solve spectrum problems, IoT objects should have cognitive capacities (Tragos and Angelakis 2013). Thus, trends are evolving toward the adaptability of intelligent or cognitive radio to the IoT. The radioelectric spectrum is generally under-used much of the time, which has created problems with spectrum scarcity. The

Intelligent Security Management and Control in the IoT,
coordinated by Mohamed-Aymen CHALOUF.

under-used spectrum is called a spectral hole or white space (Khan *et al.* 2016). To profit from these white spaces, Cognitive Radio (or CR) is becoming essential and promising in various application domains such as intelligent networks (Zhang and Chen 2012; Amraoui *et al.* 2014) and Cognitive Radio Sensor Networks (or CRSN) (Lin *et al.* 2013; Rawat *et al.* 2016).

Cognitive radio nodes are intelligent wireless devices endowed with cognitive capacities for searching for holes in the spectrum and taking opportunist decisions so as to transmit without interference with legitimate users. In intelligent radio, radio systems have the capacity to take stock of their environment via sensors and can react to the information that these provide.

The main contributions of this chapter are the following:

– presenting a complete study on the challenges and enabling technologies of the IoT;

– presenting a complete study on the motivations and challenges of applying application radio in the IoT;

– categorizing the articles studied according to their challenges and the type of network as well as the validation tool used.

The remainder of the chapter is organized in the following manner. Section 3.2 presents the IoT: its definition, applications and challenges as well as enabling technologies. Section 3.3 presents intelligent radio; we are clearly interested in its motivations and application challenges in the IoT. Finally, section 3.4 is reserved for the conclusion.

3.2. Internet of Things (IoT)

3.2.1. *Definition of the IoT*

According to Perera *et al.* (2013), "the IoT is a global network of interconnected objects that can be addressed in a unique way, based on standard communication protocols, and it enables individuals and things to be connected at any moment, anywhere, with anything and anyone, ideally by using any network and any service".

As Figure 3.1 shows, the IoT has four specific visions: things-oriented, Internet-oriented, semantic-oriented and tactile-oriented. The motivation behind the things-oriented vision is the monitoring of objects with the help of a network based on omnipresent sensors and technologies using RFID, since each object can be identified uniquely by its Electronic Product Code (or EPC). For the Internet-oriented vision, it has become more useful thanks to the IP network already developed.

The semantic-oriented vision comes into play for processing the raw data generated by sensor-based systems with a separation between the data and their interpretation. The environments and semantic execution architectures handle problems linked to representation, storage, interconnection, and the recovery and organization of information generated by the IoT (Atzori *et al.* 2010).

The tactile-oriented vision is the most recent vision of the IoT in which the haptics, or touch and the senses, can be communicated from one geographic location to another, making it possible to control and navigate distant objects almost in real time.

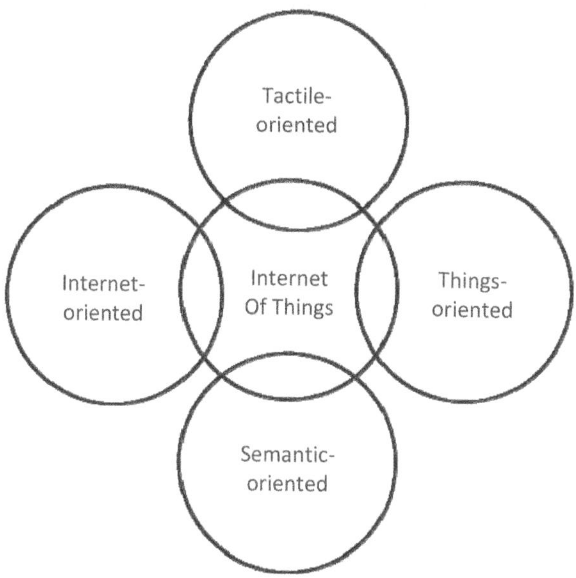

Figure 3.1. *The four visions of the IoT (Akinyoade and Eluwole 2019)*

3.2.2. Applications of the IoT

IoT objects are supposed to be usable in several domains. In what follows, we cite their most significant application domains:

– *Smart cities*: the vision of the smart city is based on information and communication technologies to provide services and production infrastructures linked to ecosystems (Somov *et al.* 2013). There are many domains, such as smart transport, smart security and smart energy management. Some applications current in the smart city are waste management, indoor positioning systems, management of parking spaces, street/traffic lighting, crowd movement analysis, etc. (Borgia 2014; Miz and Hahanov 2014; Perera *et al.* 2014).

– *Applications in the home*: IoT applications in the home include objects or devices currently found in our houses (Tragos and Angelakis 2013). Door locks using GPS and personal ID to open or close doors, the smart thermostat to control cooling or heating remotely and smart scales for monitoring weight, heart rate and temperature are only some of these current examples.

– *Environmental applications*: environment monitoring, air quality monitoring, public sensor infrastructures, smart agriculture, etc., are current examples of this type of application (Tragos and Angelakis 2013).

– *Health and personal monitoring*: IoT applications act as miniature smart sensors. These sensors are placed on or near the object to be monitored or controlled. Typical applications are remote health monitoring, physical condition monitoring, sports watches, baby monitors, etc. (Borgia 2014).

– *Industry*: applications in this domain are linked largely to manufacturing processes, such as fire safety, supply chain management and the management of manufacturing processes (Borgia 2014). The most important application in this domain is RFID object-tagging. These RFID-tagged objects need a chip and a wireless communication method for purchases, production, identification, monitoring, distribution and transport.

3.2.3. *IoT challenges*

The IoT uses different technologies, such as Cloud services, data modeling, storage, reasoning, sensors, processing and communication technologies to build a global vision (Perera *et al.* 2013). Communication mechanisms in the IoT are meant to be wired or wireless (WiFi, satellite, etc.). Wireless sensor networks (or WSN) or Wireless Mesh Networks (or WMN) can act as potential wireless solutions; however, spectrum availability limits their integration into the IoT. Communication methods should be secure, fast, reliable, uniquely addressable and interoperable before they can be widely adopted. And all this should be assured by a smart, transparent and autonomous functioning of objects in the network (Atzori *et al.* 2010).

The different challenges of the IoT are present below (Alam *et al.* 2020):

– *Interoperability*: in the IoT, there will be large devices with heterogeneous technologies involved, and there should be a mechanism for this so that these technologies and devices operate and communicate with one another effectively.

– *Scaling up*: the IoT network has billions of devices and is developing rapidly with more and more solutions for normalization and security in this domain. The network should be evolutive for any number of devices without affecting performances. The IoT should be able to evolve over time and adapt to new technologies.

– *Availability*: the IoT network requires availability 24/7 to meet the increase in user demand.

– *Performance*: the system based on the IoT should compromise between low power and high performances. It should offer high performances with low energy.

– *Resilience*: resilience refers to the system's capacity to recover from any possible failure, to resist external threats and to adapt to changes in the environments in which it operates.

– *Security and confidentiality*: the most important requirement in any network is security and confidentiality. This is becoming more critical because of the networking of many resource constrained devices where traditional security and encryption cannot be applied and a completely different architecture is required to guarantee the security and confidentiality of these networks.

3.2.4. *Enabling technologies in the IoT*

In this section, we will discuss the enabling technologies adopted in literature for the IoT.

3.2.4.1. Cloud/Fog/Edge Computing

Cloud Computing meets user requirements such as providing services on demand and managing access and omnipresent resources (Biswas and Giaffreda 2014; Borgia 2014). It was envisaged that the convergence of Cloud Computing and the IoT would have two main approaches: Cloud-based IoT and IoT-centric Cloud. Cloud-based IoT will provide IoT features in the Cloud, whereas the IoT-centric Cloud will provide Cloud features in the IoT (Biswas and Giaffreda 2014).

Fog Computing defines a combination of the traditional centralized data storage system and the Cloud. The calculation is carried out on local networks, although the servers themselves are decentralized.

As for Edge Computing, it processes data remotely from centralized storage, while still conserving information on local parts of the network (an edge device). When the data are sent to the peripheral device, they can be processed on it directly, without being sent to the centralized Cloud.

Figure 3.2 shows the difference between Cloud, Fog and Edge.

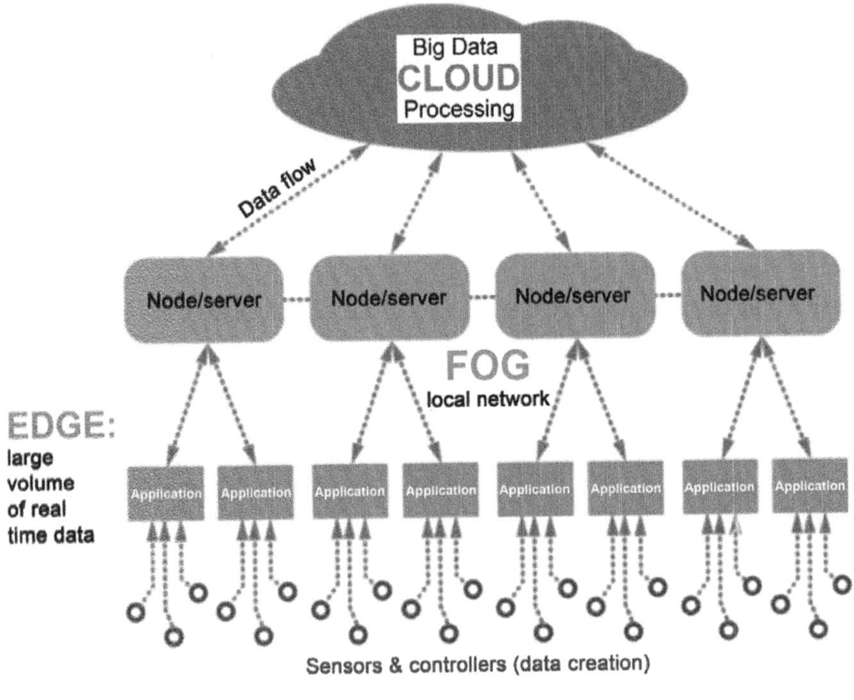

Figure 3.2. *Cloud, Fog and Edge. For a color version of this figure, see www.iste.co.uk/chalouf/intelligent.zip*

Connected objects can generate data that will be analyzed in real time at edge-level. Data that are critical at this level can be sent to the Cloud for global analysis and massive data treatment. This is typical for self-driving cars equipped with connected objects and a super calculator to gain response time via edge-level processing.

3.2.4.2. *Network of wireless sensors*

IEEE 802.15.4 is a standard that supports a low binary flow and operates at a low power and it is broadly used in wireless network sensors. Several solutions have been suggested in this domain, such as the WISP (Wireless Identification and Sensing

Platforms) project carried out at Intel Labs (Atzori *et al.* 2010). WISP is used in some environments on standard RFID readers to measure quantities such as temperature, acceleration, light, distortion and the level of liquid.

3.2.4.3. *5G*

5G will be a key element of the networked society. It will support a large number of connected devices and will meet communication needs, very reliably, in real time for critical applications. 5G will provide wireless connectivity for a broad range of new applications and usage cases, including portable devices, smart homes, security/traffic control, critical infrastructures and industrial processes. Consequently, this will equally accelerate the development of the IoT. 5G technology will become a key driver for the IoT. The main characteristics of 5G include very high data flows (generally expressed in Gbps), an extremely low latency, an enormous increase in the capacity of the base station and a significant improvement in the quality of service (QoS) perceived by users (Goudos *et al.* 2017).

3.2.4.4. *LPWA* (Low Power Wide Area) *technologies for the mobile IoT*

LPWA technologies support low power use, low data needs and long range operating devices. This notably involves a solution belonging to LoRa Alliance, LoRaWAN (LoRa Alliance 2015) and current and future cellular technologies. LoRaWAN defines the system's communication protocol and architecture for the network, whereas the LoRa physical layer provides a long-range communication link. LoRaWAN operates on unlicensed frequency bands lower than 1 GHz, which are different for each region of the world (Goudos *et al.* 2017).

3.2.4.5. *Big Data*

The term Big Data designates information with three main characteristics: speed, variety and volume. Speed refers to the data processing speed. However, the main characteristic of Big Data is the chance to analyze it in order to extract usable information. As the number of connected objects is increasing, the volume of data generated by the IOT is exploding. Thus, to be able to take stock of it and analyze it in real time, it is necessary to rely on Big Data analytical tools. These tools have the capacity to process, quickly, the large volumes of data generated continuously by IoT devices and extract usable information. In 2020 and according to Gartner (IT service management company), 20.8 billion connected objects have been used in the world. On the other hand, IDC (International Data Corporation) shows that the IoT has extended to 212 billion devices in the world. The resulting increase in the volume of data will offer new opportunities, but will also present obstacles to be overcome.

3.2.4.6. *Machine learning*

The IoT needs algorithms capable of analyzing diverse data sources in real time. New attempts have been made to solve this problem. For example, deep learning algorithms, which are a form of neural networks incorporating the evolution, can reach great precision if they have enough data and time. However, algorithms based on neural networks lack interpretation, that is, scientists cannot understand the reasons behind the results of the model. To do this, semi-supervised algorithms, that model a small quantity of tagged data with a large quantity of untagged data, can help with analysis of IoT data (Mahdavinejad *et al.* 2018).

3.2.4.7. *Communication protocols*

At different levels, the IoT uses communication protocols different from those used on an ordinary machine on the Internet. There are several options for the physical layer, such as RFID, Bluetooth Low-Energy (BLE), IEEE 802.11ah and Near Field Communication (NFC), whereas for the MAC layer, the most current option is IEEE 802.15.4. The network layer requires use of an adaptation protocol such as 6LoWPAN to compress and fragment IPv6 headers. In the transport layer, UDP is used, whereas in the application layer, a constrained version of HTTP CoAP (Constrained Application Protocol) is used. The data layer uses the compressed XML format, the EXI (Efficient XML Interchange) protocol (Goudos *et al.* 2017).

3.3. Intelligent radio

3.3.1. *Definition of intelligent radio*

Intelligent or cognitive (RC) radio exploits opportunistic access to frequency bands by Secondary Users (SUs, also called unlicensed users) to increase spectral efficiency. When Primary Users (or PUs, also called licensed users) do not transmit, white spaces or spectrum deviations are created in the spectrum. Without interfering with the PUs, the SUs access these spectrum holes opportunistically using the white spaces belonging to PUs (Benmammar and Amraoui 2013; Amraoui and Benmammar 2015).

The cognitive radio system needs four main functions that enable it to use the spectrum opportunistically (Akyildiz *et al.* 2009). These functions use the four stages of the RC terminal to manage the spectrum, which are spectrum detection, spectrum decision, spectrum analysis and spectrum mobility.

3.3.1.1. *Spectrum detection*

Spectrum detection is the basic feature of intelligent radio; it consists of detecting the unused spectrum and of sharing it without interference with other users. One of the aims of spectrum detection, in particular for the detection of interferences, is to obtain the spectrum state (free or occupied), for the spectrum to be accessible to the SU under interference constraint.

3.3.1.2. *Spectrum decision*

A decision model is needed to access the spectrum. The complexity of this model depends on the parameters considered in spectrum analysis. The decision becomes more complex when an SU has several objectives. For example, an SU can be intended to maximize performances while still minimizing the disturbances caused to the PU. Stochastic optimization methods will be a useful tool for modeling and solving the problem of spectrum access in intelligent radio.

When several users (primary and secondary) are in the system, preference will influence the decision to access the spectrum. These users may or may not be very co-operative in accessing the spectrum. In an uncooperative environment, each user has their own objective, whereas in a cooperative environment, all users can work together to reach an objective.

3.3.1.3. *Spectrum analysis*

The results of spectrum detection are analyzed to estimate the quality of the spectrum. This quality can be characterized in relation to the Signal/Noise Ratio (or SNR), the average correlation and the availability of white spaces. Information on the quality of the spectrum available for an intelligent radio user can be imprecise and noisy. Learning algorithms from the domain of artificial intelligence can be used by users to analyze the spectrum.

3.3.1.4. *Spectrum mobility*

Spectrum mobility is the process that allows the user of intelligent radio to change its operating frequency. RRC cognitive radio networks make it possible to use the spectrum dynamically, allowing radio terminals to operate in the best available frequency band, to maintain a transparent communication during the transition to a better frequency.

Figure 3.3 illustrates the four main spectrum functions for managing the cognitive radio cycle as well as the possible transitions between them.

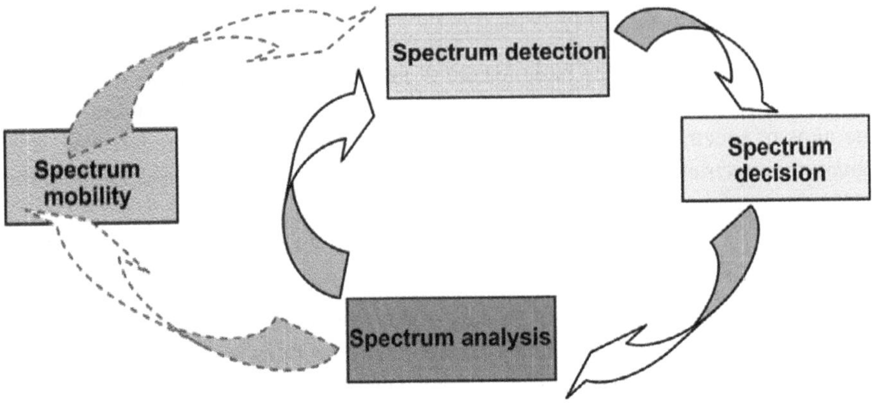

Figure 3.3. *Features for spectrum management in intelligent radio. For a color version of this figure, see www.iste.co.uk/chalouf/intelligent.zip*

Intelligent radio would also open new horizons in emerging domains such as intelligent technology, satellite communications, defense, public security, health monitoring, next generation technologies and the IoT (Doctor and Iqbal 2012; Mahmud *et al.* 2016).

3.3.2. *Motivations for using intelligent radio in the IoT*

Recently, most research on the IoT is concentrated on communication, data storage and connectivity. However, without full cognitive capacity, the IoT will not be able to overcome mounting difficulties.

Two modes of cognitive communication are considered essential for the IoT: the first is intra-cognitive communication and the second is inter-cognitive communication. In intra mode, the same cognitive capacities are attributed to cognitive objects, whereas in inter mode, objects have different cognitive capacities (Wu *et al.* 2014).

It has been envisaged that IoT future objects should have cognitive capacities to consider, learn and understand the physical and social world themselves (Shah *et al.* 2013; Wu *et al.* 2014). This means that these objects should have a perception action cycle, intelligent decision-making, knowledge discovery, data analysis and the provision of services on demand. Consequently, the integration of intelligent radio in the IoT should become a priority.

Intelligent radio is a promising solution for the IoT for various reasons:

– For a large number of IoT objects, the situation will be aggravated, since it will not be possible to allocate bandwidth to this large number of devices. At the same time, the number of licensed users will also increase, creating problems for unlicensed users (secondary or opportunistic). The cost of purchasing bandwidth will also be high. This entails solutions that are ready to be deployed, and intelligent radio may be one of these options.

– The increasing number of IoT objects will suffer problems of interference when objects move from one place to another. IoT objects based on intelligent radio can opportunistically seek channels without interference for better communication.

– Structures based on the IoT predict a situation where individuals possessing accessible objects on the Internet will be constantly in motion. Continuous connectivity to these objects will suffer as a result since a single wireless mechanism may not be available everywhere. Intelligent objects endowed with cognitive capacities can provide connectivity on the move.

– In the future, people will want to use objects without worrying about data storage and connectivity. Machines endowed with cognitive capacities will independently seek storage space and will continuously detect the spectrum to move data toward Cloud servers.

– Traditional wireless communication technologies do not make it possible to share the spectrum between several users. Future cellular networks will also suffer from problems with spectrum sharing. Intelligent radio presents this advantage and, in the future, it will be beneficial when there are a number of machines, each seeking to access the spectrum. However, regulatory, commercial and technological frameworks should be envisaged for implementing spectrum sharing zones.

Figure 3.4 shows the integration of several technologies into the IoT, including intelligent radio, where many problems need to be resolved. The main challenges, such as providing a higher level of efficiency, improving adaptability based on the existing environment, standardization and industrialization, the interoperability of interconnected devices and, above all, the guarantee of confidence and security in the omnipresent environment, should be considered long term in the growth of both technologies (Rawat *et al.* 2016).

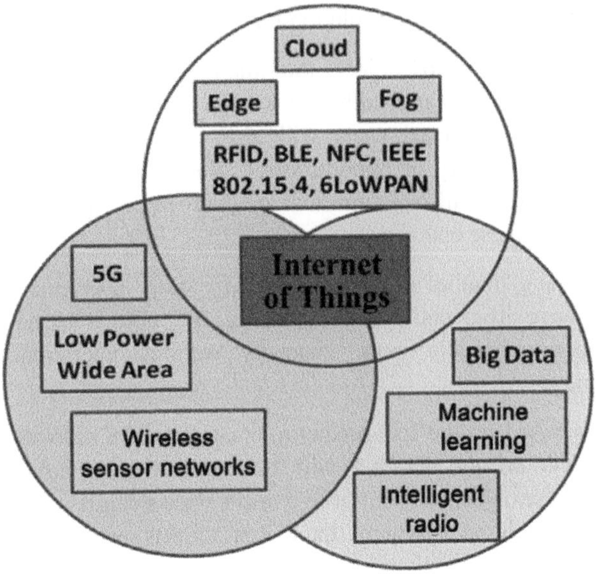

Figure 3.4. *Integration of multiple technologies in the IoT. For a color version of this figure, see www.iste.co.uk/chalouf/intelligent.zip*

Existing wireless sensor networks use ISM (industrial, scientific and medical) bands for communication (Wu *et al.* 2017). These bands are used by many other technologies, which can lead to long waiting times, in particular for traffic sensitive to delays such as the VoIP protocol. Research has shown that this coexistence in the ISM band can reduce performances of wireless sensor networks and increase waiting times. Consequently, wireless sensor networks require an interconnection with the physical world and intelligent decision-making based on the surrounding environment with the help of intelligent radio (Alves *et al.* 2017).

3.3.3. *Challenges in using intelligent radio in the IoT*

Intelligent radio is a promising technology for the IoT to give these objects cognitive capacity. However, energy recovery, spectrum allocation and spectrum detection remain the main challenges that can prevent the deployment of intelligent radio in the IoT.

In what follows, we present some work that focuses on resolving these challenges in an IoT context-based intelligent radio.

3.3.3.1. *Energy recovery*

Intelligent radio nodes should operate while still maintaining their energy levels and, consequently, energy recovery schemas should be applied.

Most work in this domain requires the functioning of the node to be optimized to reach energy efficiency. It should be noted that for intelligent radio, several techniques are used to minimize energy consumption at the level of the MAC layer (Hesham *et al.* 2012; Wang *et al.* 2014; Khan *et al.* 2015).

The protocol proposed by the authors in Qureshi *et al.* (2017) consumes less calculation time and transmits energy at higher speed compared to cognitive radio reference MAC protocols.

The authors of Ren *et al.* (2018) have proposed a resource allocation model to demonstrate how to control, jointly, dynamic access to the spectrum and energy management to optimize the utility of the network while still ensuring its stability and durability. The authors have shown that by allowing the sensor nodes to access licensed channels opportunistically, a sensor network based on intelligent radio can provide an effective solution for managing the spectrum in the IoT.

The authors of Ozger *et al.* (2018) have shown that by combining the IoT with energy recovery techniques and intelligent radio problems with intelligent networks, such as limited battery power, can be solved.

In Aslam *et al.* (2018), the authors have a schema for energy recovery and for managing the spectrum in sensor networks based on intelligent radio. IoT architecture has been introduced, in which the sensor nodes can access the spectrum opportunistically and recover energy from ambient radiofrequency sources.

The authors of Yan *et al.* (2018) have developed a schema for designing and analyzing IoT objects assisted by cognitive energy recovery relays. An effective strategy for selecting relays in this type of network has been proposed to determine when and which relays to select to facilitate transmission.

In Alzahrani and Ejaz (2018), the authors suggested a resource management schema for the IoT based on intelligent radio with energy recovery in 5G networks. The authors' objective is to maximize flow while still guaranteeing quality of service requirements in terms of data flow and minimal residual energy constraints on each IoT node.

3.3.3.2. *Spectrum allocation*

Policies for allocating channels have been studied in intelligent radio, but their adaptations in IoT objects are mainly an unexplored domain.

The authors of Shigueta *et al.* (2014) have suggested an opportunistic allocation strategy for channels in the IoT. An algorithm based on traffic history has been suggested, guaranteeing a minimum of interferences. The proposed approach adapts quickly to changes in topology and in the channels that happen in a network.

In Zhang *et al.* (2012), the authors considered a scenario in which the PU wishes to share a fraction of its bandwidth with a single one of the SUs in exchange for a cooperative transmission when the transmission flow from the PU is lower than the QoS required.

The authors of Zheng *et al.* (2019) have proposed a clustering protocol taking into account the stability of the network for sensor networks based on intelligent radio in the IoT. Dynamic management of the spectrum and of energy consumption are simultaneously integrated into the design of the proposed protocol.

In Ali *et al.* (2018), the authors have examined the availability of a licensed channel depending on the activities of its main users (e.g. traffic models) and have proposed an admission system for calls from SUs depending on their priorities. The objective of their article is to reduce the probability of high priority SU calls being blocked while still maintaining a sufficient level of channel use.

The authors of Li *et al.* (2019) have proposed a new method for dynamic access to the spectrum based on selling the spectrum for distributed IoT devices.

In Mohapatra *et al.* (2018), the authors proposed a new asymmetrical, asynchronous channel hopping mechanism in intelligent radio based on the IoT where SUs have different sets of channels available and can enter the network without clock synchronization.

3.3.3.3. *Spectrum detection*

Operation of intelligent radio begins with spectrum detection. Over the course of this process, the cognitive radio nodes seek opportunistic access to the spectrum where PUs are present. Recently, the concept of Cloud services has had a resurgence, since multiple services are provided to all users, including infrastructure, platforms and software applications. There is also a new paradigm that consists of providing detection as a service on the IoT infrastructure (Perera *et al.* 2014).

The authors of Khan *et al.* (2017) have underlined the intelligent radio features to use in the IoT, in particular spectrum detection in conjunction with Cloud services to serve as auto-reconfigurable IoT solutions for a number of applications.

The authors of Khalid and Yu (2019) have proposed a schema for detecting spatial and temporal spectral holes in a heterogenous IoT environment based on intelligent radio. The digital results have been evaluated under different network and detection parameters to examine the sensitivities of different parameters.

To improve detection precision in CRSN, the authors of Yue *et al.* (2016) have proposed a co-operative spectrum detection schema based on secondary information. The proposed schema uses a co-operative process for spectrum detection and an effective clustering algorithm meeting the specific requirements of IoT applications.

The authors of Li *et al.* (2018) have proposed a new method of broadband spectrum detection based on the reconstruction of channel energy in the IoT using intelligent radio.

The authors of Shafiq *et al.* (2019) have proposed a new MAC protocol for multichannel IoT networks based on intelligent radio. This protocol solves not only the problem of the hidden primary terminal, but also avoids the problems of the terminal being hidden and exposed at same time by mutual spectrum detection.

The authors of Zikria *et al.* (2018) have proposed an opportunistic MAC protocol for CRSN based on the IoT. This protocol has been compared to the MAC IEEE 802.11 protocol with round robin selection schemas and random channel selection schemas.

Table 3.1 summarizes this section by giving more ample information on the work presented, especially the type of network in which IoT objects have been deployed as well as the validation or simulation tools used by the authors.

3.4. Conclusion

In this chapter, we have presented the challenges and motivations in using intelligent radio in the IoT. Without intelligence, the IoT will not be able to meet all these requirements. Indeed, with the scarcity of spectral resources, the integration of cognitive radio with the IoT has become unavoidable.

Reference	Challenges of IoT based on intelligent radio	Type of network	Validation/ simulation tools
Qureshi *et al.* (2017)	Energy recovery	*Ad hoc* network	Analytical model
Ren *et al.* (2018)	Energy recovery	Wireless sensor network	Sensor deployment in a real environment
Ozger *et al.* (2018)	Energy recovery	Smart grid	No simulation
Aslam *et al.* (2018)	Energy recovery	Wireless sensor network	Simulation
Yan *et al.* (2018)	Energy recovery	*Ad hoc* network	Analytical model
Alzahrani and Ejaz (2018)	Energy recovery	5G	Simulation with NS3
Shigueta *et al.* (2014)	Spectrum allocation	*Ad hoc* network	Simulation
Zhang *et al.* (2012)	Spectrum allocation	*Ad hoc* network	Simulation
Zheng *et al.* (2019)	Spectrum allocation	Wireless sensor network	Simulation
Ali *et al.* (2018)	Spectrum allocation	Cellular network	Simulation
Li *et al.* (2019)	Spectrum allocation	*Ad hoc* network	Simulation
Mohapatra *et al.* (2018)	Spectrum allocation	*Ad hoc* network	Simulation with OMNeT++
Khan *et al.* (2017)	Spectrum detection	*Ad hoc* network	No simulation
Khalid and Yu (2019)	Spectrum detection	Cellular network	Simulation
Yue *et al.* (2016)	Spectrum detection	Wireless sensor network	Simulation
Zikria *et al.* (2018)	Spectrum detection	*Ad hoc* network	Simulation
Shafiq *et al.* (2019)	Spectrum detection	*Ad hoc* network	Simulation with MATLAB
Li *et al.* (2018)	Spectrum detection	Wireless sensor network	Simulation with NS2

Table 3.1. *Summary of articles presented*

However, few publications focus on intelligent radio in an IoT context. Manufacturers, for their part, also need to make more effort at normalization in this domain, since we believe that intelligent radio is the key solution for the IoT to provide more autonomy, lifespan and adaptability to connected objects.

3.5. References

Aijaz, A. and Aghvami, A.H. (2015). Cognitive machine-to-machine communications for Internet-of-Things: A protocol stack perspective. *IEEE Internet of Things Journal*, 2(2), 103–112.

Akinyoade, A.J. and Eluwole, O.T. (2019). The Internet of Things: Definition, tactile-oriented vision, challenges and future research directions. In *3rd International Congress on Information and Communication Technology*. ICICT, London.

Akyildiz, I.F., Lee, W.Y., Chowdhury, K.R. (2009). CRAHNs: Cognitive radio *ad hoc* networks. *Ad Hoc Networks*, 7(5), 810–836.

Alam, S., Siddiqui, S.T., Ahmad, A., Ahmad, R., Shuaib, M. (2020). Internet of Things (IoT) enabling technologies, requirements, and security challenges. In *Advances in Data and Information Sciences*, Kolhe, M.L., Tiwari, S., Trivedi, M.C., Mishra, K.K. (eds). Springer, Singapore.

Ali, A., Feng, L., Bashir, A.K., El-Sappagh, S.H.A., Ahmed, S.H., Iqbal, M., Raja, G. (2018). Quality of service provisioning for heterogeneous services in cognitive radio-enabled Internet of Things. *IEEE Transactions on Network Science and Engineering*, 7(1), 328–342.

Alves, M.M., Pirmez, L., Rossetto, S., Delicatoa, F.C., de Farias, C.M., Pires, P.F., dos Santos, I.L., Zomaya, A.Y. (2017). Damage prediction for wind turbines using wireless sensor and actuator networks. *Journal of Network and Computer Applications*, 80, 123–140.

Alzahrani, B. and Ejaz, W. (2018). Resource management for cognitive IoT systems with RF energy harvesting in smart cities. *IEEE Access*, 6, 62717–62727.

Amraoui, A. and Benmammar, B. (2015). Dynamic spectrum access techniques: State of the art. In *Research Methods: Concepts, Methodologies, Tools, and Applications*, Information Resources Management Association (ed.). IGI Global, Hershey.

Amraoui, A., Benmammar, B., Krief, F. (2014). Cognitive radio resource management using multi-agent systems, auctions and game theory. *WSEAS Transactions on Computers*, 13(41), 463–475.

Aslam, S., Ejaz, W., Ibnkahla, M. (2018). Energy and spectral efficient cognitive radio sensor networks for Internet of Things. *IEEE Internet of Things Journal*, 5(4), 3220–3233.

Atzori, L., Iera, A., Morabito, G. (2010). The internet of things: A survey. *Computer Networks*, 54(15), 2787–2805.

Benmammar, B. and Amraoui, A. (2013). *Radio Resource Allocation and Dynamic Spectrum Access*. ISTE Ltd, London, Wiley, New York.

Biswas, A.R. and Giaffreda, R. (2014). IoT and cloud convergence: Opportunities and challenges. In *2014 IEEE World Forum on Internet of Things*. IEEE, Seoul.

Borgia, E. (2014). The Internet of Things vision: Key features, applications and open issues. *Computer Communications*, 54, 1–31.

Doctor, F. and Iqbal, R. (2012). An intelligent framework for monitoring student performance using fuzzy rule-based linguistic summarisation. In *International Conference on Fuzzy Systems*. IEEE, Brisbane.

Goudos, S.K., Dallas, P.I., Chatziefthymiou, S., Kyriazakos, S. (2017). A survey of IoT key enabling and future technologies: 5G, mobile IoT, sematic web and applications. *Wireless Personal Communications*, 97(2), 1645–1675.

Hesham, L., Sultan, A., Nafie, M., Digham, F. (2012). Distributed spectrum sensing with sequential ordered transmissions to a cognitive fusion center. *IEEE Transactions on Signal Processing*, 60(5), 2524–2538.

Khalid, W. and Yu, H. (2019). Spatial-temporal sensing and utilization in full duplex spectrum-heterogeneous cognitive radio networks for the Internet of Things. *Sensors – An Open Access Journal*, 19(6), 1–18.

Khan, D.M., Saher, N., Shahzad, F., Mohamudally, N. (2015). A relative study of load balancing and management techniques in a distributed system: The IUB case study. *Science International*, 27(4), 3171–3177.

Khan, A.A., Rehmani, M.H., Rachedi, A. (2016). When cognitive radio meets the Internet of Things? In *12th International Wireless Communications and Mobile Computing Conference (IWCMC 2016)*. IEEE, Paphos.

Khan, A.A., Rehmani, M.H., Rachedi, A. (2017). Cognitive-radio-based Internet of Things: Applications, architectures, spectrum related functionalities, and future research directions. *IEEE Wireless Communications*, 24(3), 17–25.

Li, Z., Chang, B., Wang, S., Liu, A., Zeng, F., Luo, G. (2018). Dynamic compressive wide-band spectrum sensing based on channel energy reconstruction in cognitive Internet of Things. *IEEE Transactions on Industrial Informatics*, 14(6), 2598–2607.

Li, F., Lam, K.Y., Meng, L., Luo, H., Wang, L. (2019). Trading-based dynamic spectrum access and allocation in cognitive Internet of Things. *IEEE Access*, 7, 125952–125959.

Lin, Y.H., Wang, Q., Wang, J.S., Shao, L., Tang, J. (2013). Wireless IoT platform based on SDR technology. In *2013 IEEE International Conference on Green Computing and Communications and IEEE Internet of Things and IEEE Cyber, Physical and Social Computing (CPSCom 2013)*. IEEE, Beijing.

LoRa Alliance (2015). A technical overview of LoRa and LoRaWAN. White paper, LoRa Alliance, Fremont.

Mahdavinejad, M.S., Rezvan, M., Barekatain, M., Adibi, P., Barnaghi, P., Sheth, A.P. (2018). Machine learning for Internet of Things data analysis: A survey. *Digital Communications and Networks*, 4(3), 161–175.

Mahmud, S., Iqbal, R., Doctor, F. (2016). Cloud enabled data analytics and visualization framework for health-shocks prediction. *Future Generation Computer Systems*, 65, 169–181.

Miz, V. and Hahanov, V. (2014). Smart traffic light in terms of the cognitive road traffic management system (CTMS) based on the Internet of Things. In *East-West Design & Test Symposium (EWDTS 2014)*. IEEE, Kiev.

Mohapatra, S., Sahoo, P.K., Sheu, J.-P. (2018). Spectrum allocation with guaranteed rendezvous in asynchronous cognitive radio networks for Internet of Things. *IEEE Internet of Things Journal*, 6(4), 6104–6116.

Ozger, M., Cetinkaya, O., Akan, O.B. (2018). Energy harvesting cognitive radio networking for IoT-enabled smart grid. *Mobile Networks and Applications*, 23(4), 956–966.

Perera, C., Zaslavsky, A., Christen, P., Georgakopoulos, D. (2013). Context aware computing for the Internet of Things: A survey. *IEEE Communications Surveys and Tutorials*, 16(1), 414–454.

Perera, C., Zaslavsky, A., Christen, P., Georgakopoulos, D. (2014). Sensing as a service model for smart cities supported by Internet of Things. *Transactions on Emerging Telecommunications Technologies*, 25(1), 81–93.

Qureshi, F.F., Iqbal, R., Asghar, M.N. (2017). Energy efficient wireless communication technique based on cognitive radio for Internet of Things. *Journal of Network and Computer Applications*, 89, 14–25.

Rawat, P., Singh, K.D., Bonnin, J.M. (2016). Cognitive radio for M2M and Internet of Things: A survey. *Computer Communications*, 94, 1–29.

Ren, J., Hu, J., Zhang, D., Guo, H., Zhang, Y., Shen, X. (2018). RF energy harvesting and transfer in cognitive radio sensor networks: Opportunities and challenges. *IEEE Communications Magazine*, 56(1), 104–110.

Shafiq, M., Singh, P., Ashraf, I., Ahmad, M., Ali, A., Irshad, A., Afzal, M.K., Choi, J.-G. (2019). Ranked sense multiple access control protocol for multichannel cognitive radio-based IoT networks. *Sensors – An Open Access Journal*, 19(7), 1–22.

Shah, M.A., Zhang, S., Maple, C. (2013). Cognitive radio networks for Internet of Things: Applications, challenges and future. In *19th International Conference on Automation and Computing*. IEEE, London.

Shigueta, R.F., Fonseca, M., Viana, A.C., Ziviani, A., Munaretto, A. (2014). A strategy for opportunistic cognitive channel allocation in wireless Internet of Things. In *2014 IFIP Wireless Days (WD)*. IEEE, Rio de Janeiro.

Somov, A., Dupont, C., Giaffreda, R. (2013). Supporting smart-city mobility with cognitive Internet of Things. In *2013 Future Network & Mobile Summit*. IEEE, Lisbon.

Tragos, E.Z. and Angelakis, V. (2013). Cognitive radio inspired M2M communications. In *16th International Symposium on Wireless Personal Multimedia Communications (WPMC 2013)*. IEEE, Atlantic City.

Wang, C.-X., Haider, F., Gao, X., You, X.-H., Yang, Y., Yuan, D., Aggoune, H.M., Haas, H., Fletcher, S., Hepsaydir, E. (2014). Cellular architecture and key technologies for 5G wireless communication networks. *IEEE Communications Magazine*, 52(2), 122–130.

Wu, Q., Ding, G., Xu, Y., Feng, S., Du, Z., Wang, J., Long, K. (2014). Cognitive Internet of Things: A new paradigm beyond connection. *IEEE Internet of Things Journal*, 1(2), 129–143.

Wu, F., Xu, L., Kumari, S., Li, X., Shen, J., Choo, K.K.R., Wazid, M., Das, A.K. (2017). An efficient authentication and key agreement scheme for multi-gateway wireless sensor networks in IoT deployment. *Journal of Network and Computer Applications*, 89, 72–85.

Yan, Z., Chen, S., Zhang, X., Liu, H.L. (2018). Outage performance analysis of wireless energy harvesting relay-assisted random underlay cognitive networks. *IEEE Internet of Things Journal*, 5(4), 2691–2699.

Yue, W., Wu, C., Chen, Z. (2016). Cooperative spectrum sensing based on side information for cognitive radio sensor networks in Internet of Things applications. *International Journal of Internet Protocol Technology*, 9(2/3), 121–128.

Zhang, B. and Chen, K. (2012). Selective spectrum leasing in Internet of Things via Nash bargaining solutions. In *2012 IEEE 2nd International Conference on Cloud Computing and Intelligence Systems*. IEEE, Hangzhou.

Zhang, M., Zhao, H., Zheng, R., Wu, Q., Wei, W. (2012). Cognitive Internet of Things: Concepts and application example. *International Journal of Computer Science Issues (IJCSI)*, 9(6), 151–158.

Zheng, M., Chen, S., Liang, W., Song, M. (2019). NSAC: A novel clustering protocol in cognitive radio sensor networks for Internet of Things. *IEEE Internet of Things Journal*, 6(3), 5864–5865.

Zikria, Y.B., Ishmanov, F., Afzal, M.K., Kim, S.W., Nam, S.Y., Yu, H. (2018). Opportunistic channel selection MAC protocol for cognitive radio *ad hoc* sensor networks in the Internet of Things. *Sustainable Computing: Informatics and Systems (SUSCOM)*, 18, 112–120.

4

Optimizing the Energy Consumption of IoT Devices

Ahmad KHALIL, Nader MBAREK and Olivier TOGNI
LIB, Bourgogne Franche-Comté University, Dijon, France

4.1. Introduction

The Internet of Things (IoT) represents a new communication trend in the world of IT and telecoms. This new paradigm makes it possible to improve human quality of life while ensuring advanced and intelligent services by means of interconnection between different objects (telephones, sensors, etc.) and entities (humans, animals, etc.) used in our daily life. This new evolution in the communication of information creates different challenges, to which responses are expected to improve users' Quality of Experience (QoE). One of the major challenges of the IoT is energy consumption, given that connected objects within this type of environment may have limitations in terms of energy capacity. These limitations stem from the fact that the objects are supplied via batteries and it becomes difficult, indeed impossible, to change them manually given the large number of objects that operate in this type of environment. Consequently, it is necessary to minimize the energy consumption of connected objects, making it possible to extend the lifespan of IoT systems. In this chapter, we present in section 4.2 the general concepts relating to energy optimization by defining key terms and the different techniques used. Then, we detail in section 4.3 the characteristics of energy optimization in the IoT, as well as the challenges relating to this type of optimization through existing publications and research projects in the domain. We specify in section 4.4 our framework for ensuring optimization of energy consumption autonomously for the IoT before proposing in section 4.5 our method of energy optimization integrated into this

Intelligent Security Management and Control in the IoT,
coordinated by Mohamed-Aymen CHALOUF. © ISTE Ltd 2022.

framework. Finally, section 4.6 presents conclusions and perspectives concerning the domain of energy consumption optimization in an IoT environment.

4.2. Energy optimization

Energy optimization includes the different techniques and methods used to extend a system's lifespan while still minimizing the energy consumption of its components. In what follows, we define not only the notion of a system's lifespan, but also the concepts relating to energy optimization through different studies carried out in this domain.

4.2.1. *Definitions*

4.2.1.1. *Energy optimization*

Energy optimization is the process of achieving the best efficiency of an energy system. In other words, energy optimization makes it possible to use energy resources effectively to improve system performances. Energy optimization should also take account of other criteria. In fact, while still improving the use of energy resources, the system should also be capable of responding to requirements in terms of quality of service (QoS) and security that we can regroup with the concept of level of service.

4.2.1.2. *System lifespan*

An IoT system can only fulfill its mission to offer service when it is still "living" or operational. In this context, the system's lifespan indicates its maximal utility. Consequently, the system's lifespan defines the operational time during which the system is able to execute the tasks that are incumbent on it. In the context of sensor networks with characteristics similar to the objects of the sensing layer of the IoT architecture, some research deems that the lifespan of sensor networks should be measured until the first sensor or group of sensors runs out of energy. For these publications, losing a sensor or a group of sensors causes of a loss of some system functions and consequently of all the operations in this system. Other publications argue that the lifespan of the sensor network is defined as being the maximum duration during which the deployed sensors that remain living have the capacity to monitor the phenomena of interest (IGI Global 2019).

In the context of the IoT environment, we consider in our research that the system formed by IoT objects is operational (i.e. living) when the objects deployed are able to report precise data. In other words, we believe that the IoT system operates fully when it is able to resend precise data depending on the type of IoT service offered by this system. Thus, for critical IoT services, the system is considered to be operational on condition that there are at least two active IoT objects (i.e. a non-null residual energy) able to generate precise data. In the case of non-critical IoT services, the system is considered operational on condition that there are at least two IoT objects active independently of the precision of the reported data.

4.3. Optimization techniques for energy consumption

Different techniques have been used to ensure energy optimization in different domains of IT, including the IoT, to respond to many problems. In what follows, we present some of these techniques used in many publications.

4.3.1. *The A* algorithm*

One of the most important aspects of optimizing energy consumption in the IT domain consists of minimizing this consumption when conveying information within IT infrastructures. Consequently, the choice of route taken to transport information to its destination is an essential challenge making it possible to specify the most reliable route and to consume least energy. The A* algorithm is a pathfinding algorithm in a graph between an initial node and a final node. Because of its simplicity, it is often presented as a typical example of a path planning algorithm. This algorithm was suggested for the first time by Peter Hart, Nils John Nilsson and Bertram Raphael in 1968. It is an extension of the Dijkstra algorithm (Konakalla 2014). The A* algorithm combines the functions of search at uniform cost (used in the Dijkstra algorithm to choose the minimal cost path) and heuristic search (a calculation method that quickly provides a workable solution) to effectively calculate optimal solutions. Thanks to its performances and its precision, it benefits from widespread use.

The A* algorithm is an informed search algorithm, or indeed a first-order search algorithm, which means that it is formulated in terms of weighted graphs. In fact, starting from a specific starting node in a graph, it aims to find a path toward a given target node that has the least cost (the least distance covered, the shortest time, etc.). To do this, the A* algorithm maintains a path tree coming from the starting node and extends these paths one edge at a time until its termination criterion is satisfied. At each iteration, the A* algorithm should determine which paths extend. It does so based on the cost of the path and on an estimation of the cost needed to prolong the path to the target node. More precisely, it selects the path that has the least value for equation [4.1]:

$$f(n)=g(n)+h(n) \tag{4.1}$$

In equation [4.1], n is the last node on the path traveled (the node on which the path decision has yet to be taken), $g(n)$ is the path's cost from the source to the node n and $h(n)$ is a heuristic function that estimates the cost of the cheapest path from n to the destination (Konakalla 2014).

Different publications in the literature have used the A* algorithm to optimize energy consumption in IT infrastructures. As an example, the research presented in Rana and Zaveri (2011) describes a solution based on a routing algorithm permitting choice of the optimal route with the aim of extending the lifespan of a WSN (wireless sensor network). In their proposition, the authors adapted the A* algorithm by taking into consideration a third parameter in calculating the function $f(n)$ (i.e. in addition to $g(n)$ and $h(n)$). This third parameter, called $l(n)$, keeps track of the current energy level in the node n. This parameter $l(n)$ is used with a threshold. Thus, if the residual energy from the node n reaches a value below the threshold, the algorithm should avoid this node when seeking a path. The proposed algorithm then seeks an alternative path, where no node has residual energy below the threshold. The parameter $l(n)$ makes it possible to determine the number of nodes in the current path below the energy level threshold. In this context, the cost of the path is specified by equation [4.2]. In this approach, the routing decision is taken on the basis of value of $l(n)$. If the value of $l(n)$ is identical for two nodes, then in this case the value of $f(n)$ is checked to choose the node to be added to the path and so parameters $g(n)$ and $h(n)$ will be taken into consideration:

$$f(n)=g(n)+h(n)+l(n) \tag{4.2}$$

4.3.2. *Fuzzy logic*

Fuzzy logic is a decision support tool that imitates human decision-making and extends the Boolean logic created by Lotfi Zadeh in 1965 (Dernoncourt 2011). Fuzzy logic is based on the theory of fuzzy sets by representing a body of knowledge via the degrees of condition verification. Each condition will have a degree of verification. Indeed, the state of a condition is no longer described as true or false, by a degree of accuracy. The ultimate advantage of fuzzy logic is its description of human reasoning through a set of rules announced in a natural language (Dernoncourt 2011). Fuzzy logic has been used in multiple application domains such as image stabilization, control of pH values and the choice of handover in mobile networks (Scribd 2019). Fuzzy logic calls on different important elements that make it possible to ensure an automated decision-making process. In what follows, we describe the elements of the fuzzy logic system, that is, fuzzy sub-sets, membership functions, linguistic variables, fuzzification, the basis of inference and defuzzification.

Fuzzy sub-sets make it possible to escape Boolean logic and to introduce the notion of weighted membership. This notion makes it possible to have levels in membership of an element. If X is a classic set; we call A a fuzzy sub-set of X characterized by membership function f_A:$X \rightarrow [0,1]$. This is the equivalent of the characteristic function of a classic sub-set. Thus, at each point x of X there will be an association of a real value x between 0 and 1, which represents the degree of membership of x to A. By introducing the notion of degree of membership into verification of a condition, fuzzy logic confers flexibility of reasoning, making it possible to consider imprecisions and uncertainties.

Membership functions can take different forms (sigmoid form, hyperbolic, exponential, Gaussian tangent, etc.) chosen by following expert advice or statistical studies. A membership function characterizes one or more sub-sets of an input variable. It is represented in fuzzy logic by the variable $\mu(x)$. For example, the membership function of the fuzzy sub-set A of the classical set X is represented by $\mu(x)$.

Linguistic variables in a fuzzy sub-set are formed by the triplet (X, V, (x)). X is the variable (i.e. the classic set of values), V is the variable's value range and (x) is the vocabulary chosen to define the variable X linguistically. (x) contains the fuzzy sub-sets used to characterize the classic set X. For example, three linguistic variables can be created for speed (low, medium and high), each characterized by a membership function (Dernoncourt 2011).

Fuzzification is a mechanism that makes it possible to switch from a real variable to a fuzzy variable using membership functions, making it possible to determine the degree of membership of a real value to one or more linguistic variables. The basis of inference includes different rules formulated with the help of "if" and "then". The rules are defined by the expert. There are no precise directives for establishing these rules. To apply the rules specified in the basis of inference, a decision matrix is defined by considering the fuzzy implications based on operator logics (or, and, not) and making it possible to concatenate the triggered rules.

Defuzzification makes it possible to set up a precise value from the fuzzy sets to facilitate decision-making. Several methods of defuzzification can be used. Two example, the MOM (Mean of Maxima) and the COG (Center of Gravity) methods. The MOM method defines the output as the average of the abscissa of the maxima of the fuzzy set issuing from the aggregation of output values. The COG defuzzification method defines the output as being the abscissa of the center of gravity of the surface of the membership function characterizing the fuzzy set issuing from the aggregation of the output values (Dernoncourt 2011).

Different fuzzy models exist in the literature, such as the Mamdani model (Mathworks 2019) and the Takagi-Sugeno-Kang (TSK) model (Mathworks 2019). Each of these models is specific and can be used in a particular context.

Thus, several publications used fuzzy logic models to optimize energy consumption in IT infrastructures. As an example, the work presented in Singh *et al.* (2011) aims to extend the lifespan of WSN systems. To do this, the authors use the fuzzy logic model from Mamdani to specify the direction of mobility of the base station permitting the extension of the lifespan of the corresponding WSN. Indeed, when the base station is closer to the nodes, the energy consumed to transmit the information becomes minimal.

Thus, in the fuzzy logic model used in Singh *et al.* (2011), two inputs are considered: the residual energy of the node and the distance between the node and the base station. First of all, the input data are regrouped into a number of fuzzy sets. The residual energy is represented by three fuzzy sets (*Low*, *Medium* and *High*) and the distance is also represented by three fuzzy sets (*Near*, *Medium* and *Far*). The output of the fuzzy system corresponds to a priority variable represented by five fuzzy sets (*Very Low*, *Low*, *Medium*, *High* and *Very High*). A node with low energy is a priority. The nodes distant from the base station also have high priority. In the proposed model, the base station moves toward the node that has highest priority while still following a predefined circular path. Thus, following the priority accorded to the nodes and the input values of the fuzzy logic system, the digital priority of each node is calculated. Then, a comparison between the priorities of the nodes is made to take the decision concerning direction of the movement from the base station.

4.4. Energy optimization in the IoT

4.4.1. *Characteristics of the IoT*

The IoT system relies on an architecture (see Figure 4.1) with three layers including each of the elements and well-defined entities. The lowest layer of this architecture, called device layers, includes IoT objects with constraints in terms of energy consumption as well as the high- and low-level gateways, which can be supplied continuously. At this layer of the architecture, it is necessary to optimize energy consumption, as it makes it possible to keep IoT objects alive for significant durations and consequently to make it possible to extend the lifespan of the system.

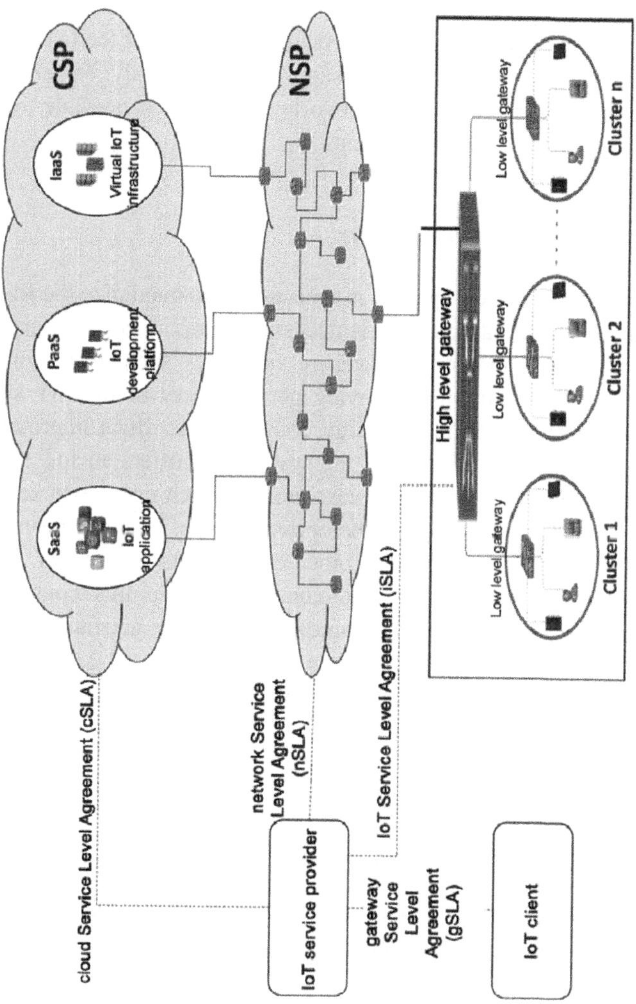

Figure 4.1. *Architecture of the Internet of Things (Khalil et al. 2018). For a color version of this figure, see www.iste.co.uk/chalouf/intelligent.zip*

The second layer is the network layer and includes infrastructure, making it possible to connect the device layers to the third layer of the architecture, that is, the Cloud layer where Cloud services are used to process and store data reported by the device layer and routed through the network layer. Optimization of energy consumption at both the network and Cloud layers is not addressed in this chapter. This optimization does not make it possible to extend the lifespan of the IoT system, given that the devices at these two layers are supplied continuously. Nevertheless, optimization of energy consumption at these two layers makes it possible to have green infrastructures respecting the environment.

4.4.2. *Challenges in energy optimization*

To extend the lifespan of an IoT system, it is necessary to maximize the lifespan of each system component. In particular, it is necessary to optimize the lifespan of components with constraints in terms of energy capacities such as IoT objects operating with batteries. In addition, improving performances of the IoT system depends strongly on the optimization of energy consumption, which improves the lifespan of IoT objects and consequently of the system that offers an IoT service. Energy consumption management is a very vast subject which extends to software and hardware. In this chapter, we focus on the optimization of energy consumption at the level of IoT objects, given that the other components of the IoT system architecture (the gateways) can be supplied continuously. In this context, the optimization can carry several factors and parameters such as the activation time for IoT objects (an inactive object preserves its energy), the frequency of data collection (sending data consumes energy), consumption due to the use of some communication technologies, energy consumption by the microprocessor and the complexity of the algorithms and software implemented (Jaxenter 2018).

4.4.3. *Research on energy optimization in the IoT*

Different approaches have been suggested in publications to minimize energy consumption in IoT systems, based on different techniques and propositions. In what follows, we present some of this work.

The authors of Li *et al.* (2020) envisage optimizing energy consumption thanks to periodic data collection in a self-powered IoT network using a NOMA (Non Orthogonal Multiple Access) technique. The self-powering characteristics allow devices to recover energy from a PB (Power Beacon) station via renewable or kinetic energy systems. In particular, the authors consider the time allocation for WPT (Wireless Power Transfer) periods, during which the energy is transferred to objects, and the WIT (Wireless Information Transfer) periods, during which the data are transferred between devices. In this context, the energy consumption of each IoT object is broken down into three parts: the first part concerns the energy consumed by activating the object, the second part concerns the energy consumed by collecting information and the third part concerns the energy used to send information to the access point. Consequently, to support the transmission of data, the energy collected by the object over the WPT period should be equal to or higher than the energy consumed during the transmission of data to the access point in the WIT period. Thus, the authors have studied the impact of dividing time slots between the WIT and WPT periods to optimize the energy consumption of self-powered IoT systems. This study has shown that the time optimal allocated to the WPT period (called α) decreases with the increase in transmission power of the PB station. When the PB station provides a higher transmission power during the WPT period, the IoT device can collect more energy over the same period. Consequently, the time for the WPT may be reduced. Moreover, the data to be transmitted in the WIT period of time, during which the data are transferred between devices, also affect the optimal duration α. When the transmission power of the PB is fixed, more WPT time is often needed to transmit more data in the following WIT duration. Similarly, the study has shown that the distance of the PB has a very significant impact on the energy recovered. Indeed, the energy recovered decreases as the distance increases.

Another piece of research presented in Zouinkhi *et al.* (2019) focuses on the question of energy consumption in WSN based on IEEE 802.15.4 technology. This work is based on IEEE standard 802.15.4 and a network topology based on a cluster tree. This network is divided into several sub-groups, each being characterized by specific IEEE 802.15.4 parameters such as the BO (Beacon Order) and SO (Superframe Order). The authors have proposed a mathematical model for calculating energy consumption to enable the PAN (personal area network) coordinator to detect the current energy level in the sensor battery. The adaptive intervention proposed is based on a comparison between the residual energy in objects with different thresholds, which leads to the self-management of energy consumption. The authors' contribution consists of collecting data on the energy of all the objects in the network. Thus, each object transmits its data to its coordinator, which calculates the battery level, named E_R. Then, the E_R level is compared to the initial energy threshold E_t (e.g. E_{t1} is defined for the objects belonging to the object cluster 1). In the case where E_R is less than E_{t1}, new specific parameters (BO1, SO1) are then calculated by the corresponding coordinator for the object or objects in question. The same procedure is repeated with all the objects and thresholds. The four battery levels addressed by the authors' approach are L1, L2, L3 and L4, corresponding to thresholds of 250 J (L1), 200 J (L2), 150 J (L3) and 100 J (L4). Thus, calculations of BO and SO become dependent on the ER level and the E_t

level. When the energy level of the objects decreases, the configuration of BO and SO changes and consequently the inactive period of the objects increases to enable energy preservation. Simulations have shown that the energy consumption of objects with low energy levels is less substantial compared to results obtained using IEEE standard 802.15.4 and other research approaches.

The authors in Li *et al.* (2013) focus on optimizing energy consumption in the IoT domain and particularly on LTE-Advanced technology. In this context, they took into consideration some IoT streaming applications. Low-speed streaming traffic can be generated by objects over a long period, imposing strict requirements in terms of energy-saving. To manage energy consumption, LTE-A includes discontinuous reception/transmission DRX/DTX (Discontinuous Reception/ Discontinuous Transmission) mechanisms to allow devices to deactivate their radio interfaces and switch to an inactive mode in different ways. The authors are interested in optimizing the DRX/DTX mechanism. Indeed, they sought to maximize periods of object inactivity while still guaranteeing QoS, in terms of the traffic bit rate, delay and packet loss rate emanating from IoT applications. Thus, effective schemas have been proposed to optimize DRX/DTX parameters and plan the transmission of packets by objects toward the base station. The key idea behind the schemas is to choose DRX/DTX configurations while still considering QoS parameters. The schema proposed by the authors is intended to minimize the objects' activation time, by reducing the unnecessary activation of objects competing to access shared resources. In addition, this schema makes it possible to optimize the period of inactivity to reduce packet delays and at the same time to configure the objects to be in a period of inactivity if this inactive period does not lead to violation of expectations in terms of QoS parameters. In parallel, the authors have broadly proposed a packet scheduling mechanism at the base station. The overall solution proposed thus makes it possible to minimize objects' energy consumption while still improving QoS parameters (loss rate, flow, jitter, etc.) for streaming IoT applications considered in this research.

4.5. Autonomous energy optimization framework in the IoT

4.5.1. *Autonomous computing*

The large number of heterogenous connected objects using different technologies in the IoT makes it difficult, or indeed impossible, for peripherals to be configured by human administrators, hence the need to automate some computing processes (Salehie and Tahvildari 2005). In this context, the AC (Autonomic Computing) paradigm (Sinreich 2006) makes it possible to confront the growing complexity and significant operational cost of managing IoT infrastructure. The importance of autonomous computing in the IoT environment has been emphasized by different organizations such as the ITU (International Telecommunications Union – ITU Telecommunication Standardization Sector) in its document "Y.2066" (ITU-T 2014). Thus, the ITU describes the importance of integrating autonomous computing in the IoT through different functions such as self-configuration, self-optimization and self-protection. In addition, the automation of processes is proving significant since it makes it possible to minimize the risk of human errors or indeed the incoherence that can arise when managing IoT infrastructures. Autonomous computing is needed for the various IoT layers. First, at the lower layer including connected objects, communications between objects should be minimized to ensure energy conservation. Thus, the network of connected objects should be configured with care and without errors to avoid useless communications as well as packet loss. Then, at the network layer, it is important to reconfigure the network during periods of congestion to minimize packet loss and to maximize infrastructure performances, and this while, for example, routing messages differently in a multi-hop network. At the upper layer, where the Cloud is situated, hosting IoT applications and comprising calculation and storage resources, autonomous computing is needed to ensure the availability of resources thanks to a distribution of the traffic load over the totality of resources in real time depending on the requests from lower layers. Similarly, IoT applications can necessitate autonomous computing capacities that modify requirements during execution. Thus, autonomous computing will make it possible to reconfigure IoT objects to collect specific data to improve the application's effectiveness and the precision of the service rendered (Weyns *et al.* 2018). If the system judges that the recorded data are not reliable, it will be able, through autonomous computing, to reconfigure specific IoT objects to get more information or even to change the frequency with which information is recorded to improve the precision of the service rendered. Without autonomous computing at the different layers of the IoT, problems will be resolved either by oversupply of resources, enormously increasing the cost of services, or through excessive use of resources leading to inefficiencies and to a reduced lifespan of IoT systems. Autonomous computing makes it possible to avoid human intervention, which is not only very costly but also subject to error. Consequently, to meet the demands of IoT applications and to overcome continuous changes in the environment, adoption of the autonomous computing concept is vital in the IoT. The switch from a manual system calling on human intervention to a totally autonomous system happens progressively following several stages defined by the Tivoli group from IBM (Ashraf *et al.* 2014).

According to this same group, Autonomous Entities (AE) interact to manage the corresponding system autonomously. Each AE has two components: Autonomic Managers (AM) that execute rules defined by administrators and implement autonomous computing functions (self-CHOP: self-Configuring, self-Healing, self-Optimizing, self-Protecting) to monitor and control the behavior of the Managed Element (ME). This latter component may be a peripheral, a software or indeed a whole system. The ME resource includes sensors and effectors. Autonomous Computing (AC) is the entity that ensures central control and analysis of these resources through a control loop called MAPE-K (Ashraf *et al.* 2014). The MAPE-K control loop specified by IBM and represented by Figure 4.2 consists of several functions (monitor, analyze, plan, execute) as well as a component called the knowledge base.

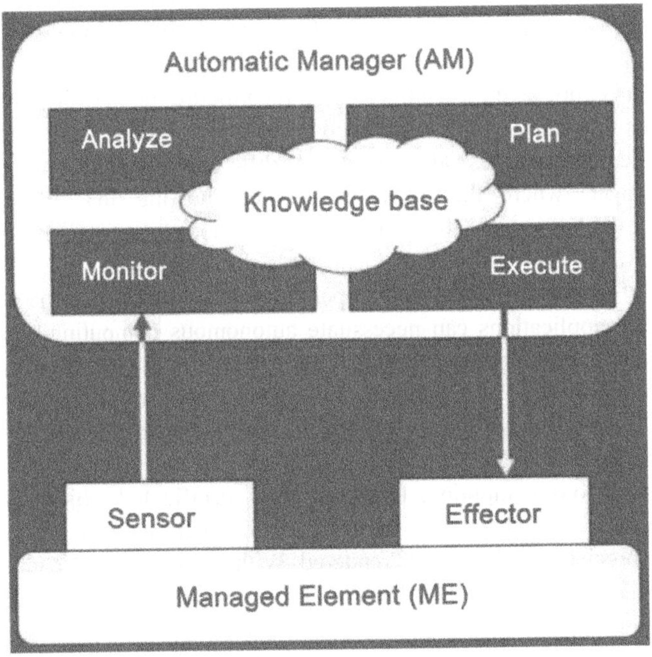

Figure 4.2. *MAPE-K Control loop*

Each function has a specific role in the closed control loop process. The "monitor" function collects information from managed resources, via sensor type interfaces. The collected data are considered by the "analyze" function to predict future system states. Subsequently, the "plan" function makes it possible to program actions depending on the strategy information and rules provided by the "knowledge base". Finally, the "execute" function controls execution of the action plan provided and distributes the recommended actions on the managed resource effectors. In this context, the knowledge stores policy information and rules corresponding to different scenarios that can be produced in the autonomous system environment (Dobson *et al.* 2006). Sensor type interfaces are used to communicate the information detected by the managed resources to the autonomous managers, while the effector interfaces are used by the autonomous managers to execute actions.

4.5.2. *Framework specification*

In the context of autonomous computing in an IoT environment, we propose a framework for the self-optimization of energy consumption, illustrated by Figure 4.3. It is based on an LL-Gw (Low-Level Gateway) from the sensing layer of our IoT architecture making it possible to collect information emanating from different IoT objects and to transmit this information to HL-Gw (High-Level Gateways) for data processing and aggregation. Low-level and high-level gateways are supplied continuously, while IoT objects have energy constraints due to the use of batteries. Thus, to extend the lifespan of the IoT system, we must consider extending the lifespan of IoT objects by minimizing their energy consumption. Our framework then aims to minimize the energy consumption of IoT objects by collecting the information from a limited number of objects selected according to an autonomous decision-making system.

The autonomous computing process in our framework for self-optimizing the energy consumption of IoT objects is implemented via implementation of the closed MAPE-K control loop at the low-level gateway (LL-Gw, considered as the AM) and the IoT objects (considered as the AE), as illustrated in Figure 4.3.

Thus, the LL-Gw, playing the role of AM, guarantees the four functions of the closed control loop (i.e. monitor, analyze, plan and execute). It also has the knowledge base. The "monitor" function of the LL-Gw makes it possible to collect, regularly, information relating to the IoT objects to supply the energy consumption self-optimization process. The "sensor" interface of the IoT object is used to record this information in real time. Then, the "analyze" function makes it possible to launch the process of evaluating the IoT objects linked to the LL-Gw based on the information collected by the "monitor" function and the information available in the knowledge base. This evaluation is based on a fuzzy logic model. Then, the "plan" function makes it possible to decide on the object(s) to activate based on the evaluation carried out by the "analyze" function and a decision-making algorithm. This decision will be implemented by the "execute" function by communicating it to the IoT objects concerned through their "effector" interfaces. Thus, this interface receives the decisions sent by the LL-Gw to configure the activity of the object and thus optimizes its energy consumption autonomously.

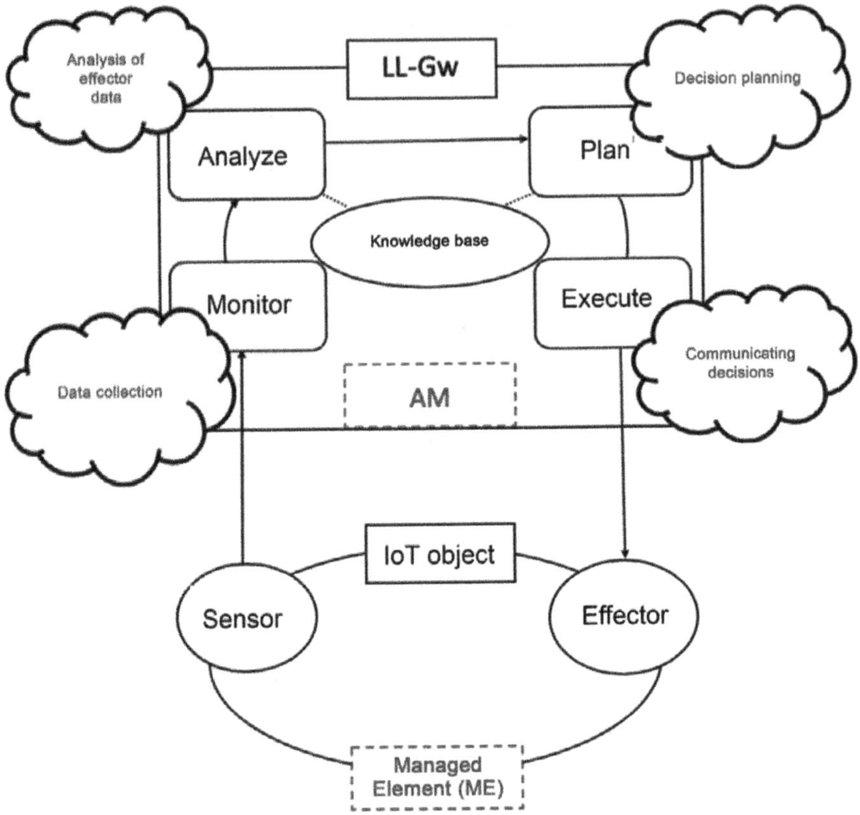

Figure 4.3. *MAPE-K loop from the IoT energy self-optimization framework. For a color version of this figure, see www.iste.co.uk/chalouf/intelligent.zip*

4.6. Proposition of a self-optimization method for energy consumption in the IoT

Our method for self-optimizing energy consumption in the IoT environment consists of a system based on fuzzy logic that will make it possible to evaluate the different IoT objects according to different criteria, as well as algorithms for selecting IoT objects to activate the collection of different types of data. In what follows, we present our fuzzy logic model making it possible to implement the "analyze" function of the MAPE-K loop to evaluate the information picked up by the IoT objects. In addition, we specify a decision-making algorithm making it possible to implement the "plan" function, which will take the decision that will be communicated to the IoT objects thanks to the "execute" function and effector interfaces.

4.6.1. *Fuzzy logic model*

In our proposition for self-optimizing the energy consumption of IoT objects, we specify a system for evaluating these objects based on the Mamdani fuzzy logic model with three input parameters (distance, percentage of residual energy and precision) and an output parameter (the chance of the IoT object) linked to different membership functions as well as rules and a basis of inference. In what follows, we detail the parameters and the basis of inference in our fuzzy logic system based on the Mamdani model, making it possible to evaluate IoT objects.

The first input parameter that we consider in our fuzzy logic system is the distance between IoT objects and the coordinator (i.e. the LL-Gw). In fact, the distance has a direct impact on energy consumption. As mentioned in Masdari and Naghiloo (2017), the energy consumption of each object depends directly on the distance that separates the coordinator. The distance between the IoT objects and the coordinator cannot exceed the range specified in IEEE standard 802.15.4. Consequently, the range of possible values that can be considered is between 0 and 100 m. In Kim *et al.* (2015), the authors studied energy function according to the distance between the coordinator and the objects for IEEE standard 802.15.4. This study has demonstrated that the energy consumption is less than 100 mJ for a distance between 0 and 30 m, between 100 and 110 mJ for a distance between 30 and 60 m, and higher than 110 mJ for a distance between 60 and 100 m. To evaluate the distance between the IoT objects and the coordinator, that is, the LL-Gw in our architecture, the LL-Gw receives the coordinates of the IoT object through the sensor interface implanted on it. Following reception of these coordinates, the LL-Gw calculates the distance using equation [4.3]:

$$\text{Distance} = \sqrt{(Xa - Xb)^2 + (Ya - Yb)^2 + (Za - Zb)^2} \qquad [4.3]$$

In this context, we use the following linguistic variables for the fuzzy set of the "distance" parameter: *Close*, *Mid* and *Far*. We specify in Figure 4.4 the trapezoid membership functions for these three linguistic variables relating to the distance input parameter. Considering the distance as an input parameter, we aim to favor activation of the IoT objects closest to the LL-Gw to minimize the energy consumption needed to transmit the transmission data.

The second input parameter that we consider in our fuzzy logic system for evaluating IoT objects is the percentage of these IoT objects' residual energy. To maximize the lifespan of the IoT system, we should optimize the use of residual energy in IoT objects. The value of this parameter corresponds to a percentage (between 0% and 100%) calculated according to equation [4.4]. In fact, the IoT object communicates, via its sensor type interface, its residual energy level at the LL-Gw, and this gateway calculates the percentage of this residual energy compared to the initial energy communicated by this object during its initialization. In this context, we use the following linguistic variables for the fuzzy set of the input parameter "percentage of residual energy": *Low*, *Mid* and *High*, given that the higher the percentage of the object's residual energy, the greater the chances of selecting this object. We specify in Figure 4.5 the triangular membership functions for these three linguistic variables relative to the "percentage of residual energy" input parameter:

$$\text{Residual energy (\%)} = \frac{Residual\ energy}{Initial\ energy} \times 100 \qquad [4.4]$$

By considering the percentage of residual energy as an input parameter, we aim to favor activation of the IoT objects with a percentage of residual energy, making it possible to prolong the lifespan of different objects in the system and, consequently, of the IoT system formed by these objects.

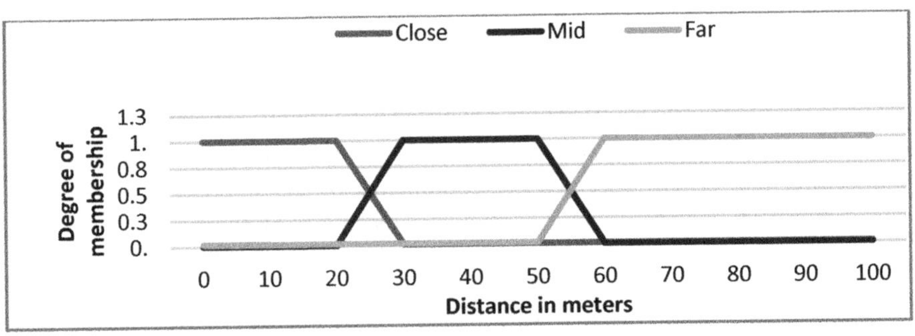

Figure 4.4. *Membership functions of the "distance" parameter. For a color version of this figure, see www.iste.co.uk/chalouf/intelligent.zip*

The third input parameter we consider in our fuzzy logic system for evaluating IoT objects is the precision of the data picked up by the object. When maximizing the lifespan of the IoT system, including the objects supplied by the batteries and the gateways supplied via an electric connection, it is necessary to consider the quality of the collected data to preserve the reliability of the corresponding IOT service.

Several criteria for quality are used to qualify data in research publications. As such, we cite precision (comparison with reference data), completeness (percentage of missing data), etc. (Baker *et al.* 2001; Wyatt and Sullivan 2005; Claudia *et al.* 2010). ISO standard 5725 (ISO 1994) uses trueness and precision to describe a method for measuring data quality. According to this standard, trueness describes the precision of the average value. This precision can be obtained by comparing the average value of a broad series of test results and an accepted reference value. In our system for evaluating IoT objects, we use data precision as an input parameter describing data quality. The value of the "precision" parameter corresponds to a percentage (between 0% and 100%) calculated according to equation [4.5]. The precision of the information picked up by each IoT object is calculated by the LL-Gw following reception of information from all objects. In this context, we use the following linguistic variables of the fuzzy set of the "data precision" input parameter: Low, Mid and High. The higher the precision of the data generated by the IoT object, the greater the chances of the object being selected. In Figure 4.6, we specify trapezoidal membership functions for these three linguistic variables relating to the input parameter "data precision". We note in Figure 4.6 that the membership function *High* reaches the maximum value of degree of membership for a precision from 97% supposing that this value has been specified in an iSLA (IoT Service Level Agreement) type contract (Khalil *et al.* 2017):

$$\text{Precision (\%)} = 1 - \frac{|Data\ value - Average\ value|}{Average\ value}) \times 100 \qquad [4.5]$$

By considering the precision of data picked up by IoT objects as an input parameter, we aim to favor activation of objects with a precision that allows some reliability of the corresponding IoT service to be maintained.

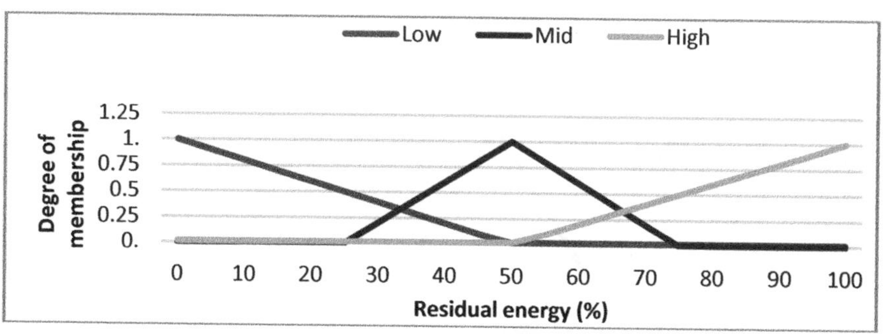

Figure 4.5. *Membership functions of the "residual energy" parameter. For a color version of this figure, see www.iste.co.uk/chalouf/intelligent.zip*

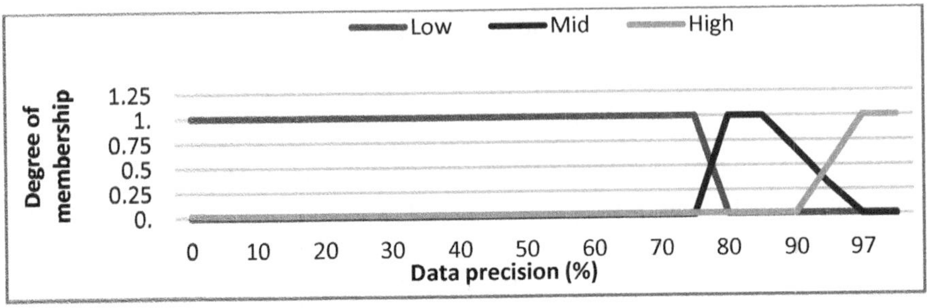

Figure 4.6. *Membership functions of the "precision" parameter. For a color version of this figure, see www.iste.co.uk/chalouf/intelligent.zip*

The output from our evaluation system based on a fuzzy logic model makes it possible to determine the IoT object's chance of selection (i.e. the chance of the IoT object) through a note in the interval [0, 10]. Indeed, following the injection of the input parameters into the fuzzy logic system, a fuzzy output is produced. The fuzzy output is characterized with the help of the output membership functions represented by Figure 4.7. Thus, nine levels are considered for the chance of the IoT object: Ultra Low, Very Low, Low, Average Low, Average, Average High, High, Very High and Ultra High. Triangular membership functions are used to characterize each level. Defuzzification makes it possible to calculate the digital value of the chance of the corresponding IoT object at the fuzzy output by determining the COG of the fuzzy zone designated by the Mamdani fuzzy logic system.

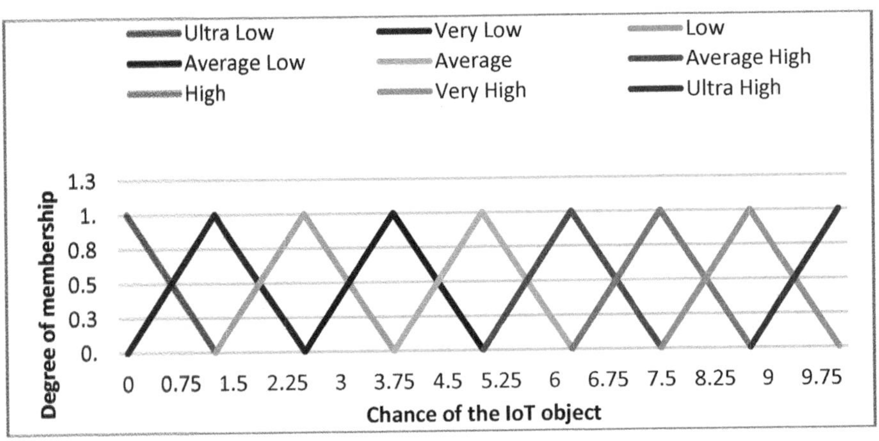

Figure 4.7. *Membership functions of the output parameter. For a color version of this figure, see www.iste.co.uk/chalouf/intelligent.zip*

To determine the system's fuzzy output, predefined rules are used from a basis of inferences. These rules are described according to an "if" and "then" conditional format. Our system for evaluating IoT objects based on fuzzy logic has three inputs with three membership functions. Consequently, our basis of inference has 27 different rules. In the context of our system making it possible to evaluate the objects that will offer critical IoT services or not, we specify two bases of inference. The first corresponds to critical IoT services by favoring the input parameter "precision", whereas the second corresponds to non-critical IoT services where all the input parameters are equivalent. The result of each rule corresponds to one of the nine fuzzy sets of the output. To validate the effectiveness of our rules, we have carried out two simulation games to adapt so as to obtain the most effective results.

4.6.2. *Decision-making algorithm*

To choose the IoT objects to activate as well as the number of objects to activate over the course of each evaluation period (energy self-optimization update interval), the LL-Gw executes an object selection algorithm depending on the type of IoT service. Thus, if the IoT service involves a critical application, the precision of the data recorded is a very important QoS parameter for this service. Consequently, the LL-Gw activates as many objects as necessary to reach the required precision. In addition, the LL-Gw should activate at least a minimal number of objects for each evaluation interval. This function is described by the algorithm from Figure 4.8.

The decision-making algorithm shows that initially the LL-Gw activate a minimal number of objects needed for this type of service, after the initialization phase during which all the objects are activated. To activate these objects, the LL-Gw configures them with the corresponding DGI (Data Generation Interval) parameter, called minimum DGI, which in all cases is inferior to the evaluation interval. Activated objects are the objects with the highest "IoT object chance" values calculated by our evaluation system. If the precision of data relating to the critical IoT service is not respected following activation of the minimal number of IoT objects (i.e. the average precision of the activated objects is lower than the precision demanded by the service), the LL-Gw proceeds to activation of an additional IoT object at the same time (by allocating the minimum DGI to an object) to reach the precision demanded by the service. The objects that have not been configured with the minimum DGI should then be deactivated and consequently configured with the maximum DGI. The configuration of objects with the maximum DGI allowing these objects to send data according to the update interval for configurations to update the input parameters of our evaluation system based on fuzzy logic.

Algorithm Energetic auto-optimization – Critical IoT services

```
Receive data from all objects ()
Mean Value (Data)
if (IoT Objects > Minimum Number of Activated IoT Objects ) then
   i ← 0
   While ( i < Number of IoT Objects) do
        RE [i] ← Retrieve Residual Energy (IoT Object[i])
        Accuracy [i] ← Retrieve Accuracy (IoT Object[i])
        Distance [i] ← Retrieve Distance (IoT Object[i])
        IoT Object Chance [i] = Fuzzy System (RE[i], Accuracy[i], Distance[i])
        i ← i + 1
   end while
   Order by IoT Object Chance (IoT Objects [], descending Order)
   i ← 0
   While (i < Minimum number of activated IoT Objects) do
        Set Data Generation Interval to minimum (IoT Object[i])
        i ← i + 1
   end while
   Mean Value (Activated IoT Objects)
   Tmp ← Accuracy (Activated IoT Objects)
   if (Tmp >= Desired Accuracy) then
        while (i < Number of IoT Objects) do
                Set Data Generation Interval to maximum (IoT Object[i])
                i ← i + 1
        end while
   else
        while (Tmp < Desired Accuracy) do
                Set Data Generation Interval to minimum (IoT Object[i])
                Mean Value (Activated IoT Objects )
                Tmp ← Accuracy (Activated IoT Objects )
                i ← i + 1
        end while
        While (i < Number of IoT Objects) do
                Set Data Generation Interval to maximum (IoT Object[i])
                i ← i + 1
        end while
   end if
end if
```

Figure 4.8. *Decision-making algorithm*

4.6.3. *Evaluating energy self-optimization in the IoT*

In the context of validating and evaluating performances of the self-optimization function specified for autonomous management of energy consumption of IoT objects, we define simulation scenarios to prove the effectiveness of this new function added to the LL-Gw level.

Characteristics of the OMNET++ simulation environment considered when evaluating energy self-optimization are described in Table 4.1.

Parameter	Value
Packet size (layer MAC)	50 bytes
Topology	Star
Frequency	2.4 Ghz
Initial energy in the object	100 J = 8.5 mAh
Interval of the fuzzy logic evaluation system	10 s
Object/gateway communication technology	QBAIoT (Khalil *et al.* 2018)

Table 4.1. *Parameters for simulation scenarios*

To evaluate performances of the self-optimization function aiming to extend the lifespan of the IoT system, we consider the following definition, already mentioned in section 4.2.1.2, for the lifespan of the IoT system with critical and non-critical services. An IoT system offering a critical service is operational on condition that there are at least two active IoT objects (i.e. a non-null residual energy) capable of generating precise data. In the case of non-critical IoT services, the system is considered operational on condition that there are at least two IoT objects active independently of the precision of the recorded data.

We carry out evaluation simulations for energy self-optimization with scenarios 1 and 2 described in Table 4.2. These scenarios correspond to four IoT objects that are located at different distances of the LL-Gw with different precision values of the recorded information. The precision value demanded by the critical IoT service considered in these scenarios is fixed at 96%.

Scenario	Precision	Distance	Min DGI	Max DGI	Energy consumption
1	96% 97% 98% 99%	80 m 60 m 40 m 20 m	0.25 s	2 s	10.1 mA 9.25 mA 8.41 mA 7.57 mA
2	99% 98% 97% 96%	80 m 60 m 40 m 20 m	0.25 s	2 s	10.1 mA 9.25 mA 8.41 mA 7.57 mA

Table 4.2. *Characteristics of simulation scenarios 1 and 2*

Figure 4.9 shows a comparison of lifespan results from the IoT system for scenarios 1 and 2 with and without energy self-optimization. We can note that energy self-optimization assures a better system lifespan duration for both scenarios compared to non-use of this function. As scenarios 1 and 2 have different characteristics, different objects are selected at each evaluation interval. The results show that we obtain a better lifespan of the IoT system in scenario 1 as the closest objects possess better data precisions. It should be noted that the decision-making algorithm has permitted selection of two objects to activate in the two scenarios during all the evaluation intervals over the course of the simulation, and this after the initialization stage where all the objects are activated, all the objects respecting the minimal required precision (see section 4.6.2).

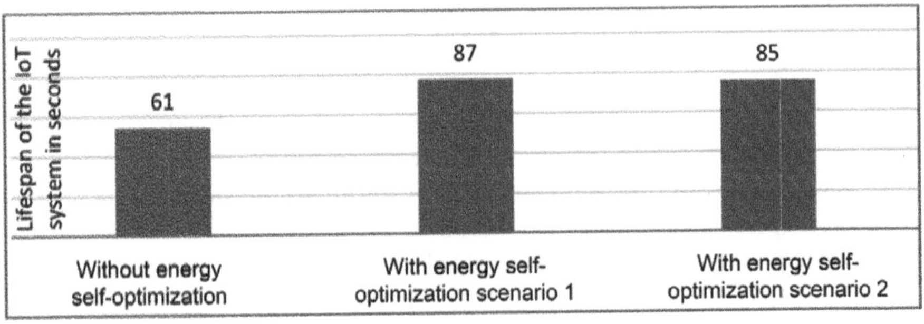

Figure 4.9. *Lifespan of the IoT system for scenarios 1 and 2*

Other simulation scenarios are presented in Table 4.3 with mobile IoT objects that change precision depending on their positions. Objects 1 and 3 from scenario 3 change position and hence precision at 25 s and objects 2 and 4 change precision at 50 s. As for scenario 4, we simulate six objects with different precisions, but at the same distance from the LL-Gw to offer a critical IoT service. Finally, the data precision required by the IoT service in scenario 3 is fixed at 95%, whereas for scenario 4 the precision required by the service is 97%.

We compare in Figure 4.10 the lifespan of the IoT system for scenario 3 with and without energy self-optimization. We can note that energy self-optimization improves this lifespan by 30%.

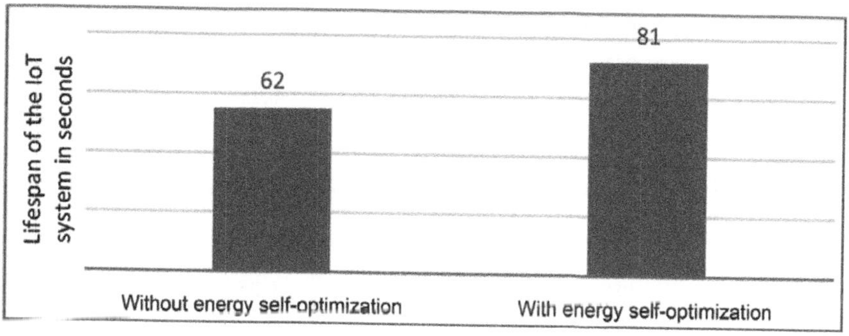

Figure 4.10. *Lifespan of the IoT system for scenario 3*

The last simulation scenario (scenario 4) has the objective of showing the effectiveness of the algorithm used in the context of energy self-optimization to select several IoT objects to respect the precision required by a critical IoT service.

In scenario 4 (see Table 4.3) the data precision emanating from some IoT objects is lower than 97%. Consequently, it is possible thanks to this algorithm (see Figure 4.8) to activate more than the minimal number of objects (i.e. two in our case of use) to recover information relating to the critical IoT service.

Figure 4.11 shows the results relating to scenario 4 on the precision of data emanating from the activated IoT objects in the use case of energy self-optimization and emanating from all the IoT objects in the case where self-optimization is not used. We can note that energy self-optimization makes it possible to obtain better data precision, except for the evaluation period between 50 s and 60 s. Indeed, at 60 s, we can note that data precision with energy self-optimization is lower than that obtained without energy self-optimization. In fact, at 60 s, three IoT objects are activated (see Figure 4.12) with the self-optimization algorithm since the precision of the objects initially activated does not meet the precision required at 97%. Consequently, an additional object has been activated to obtain a value higher than 97% whereas precision without energy self-optimization is always fixed and equal to 97.5%, as all the IoT objects are activated until their energy is spent. Thus, the lifespan of the IoT system without energy self-optimization is 67 s, whereas it is 102 s in the use case of the energy self-optimization function.

Scenario	Precision	Distance	Min/ max DGI	Energy consumption
3	99%/95% 25 s 99%/95%: 50 s 95%/99%: 25 s 95%/99%: 50 s	80 m/40 m: 25 s 80 m/40 m: 25 s 40 m/80 m: 25 s 40 m/80 m: 25 s	0.25 s/2 s	10.1 mA/8.41 mA 10.1 mA/8.41 mA 8.41 mA/10.1 mA 8.41 mA/10.1 mA
4	96% 97% 98% 99% 97% 98%	40 m	0.25 s/2 s	8.41 mA

Table 4.3. *Characteristics of simulation scenarios 3 and 4*

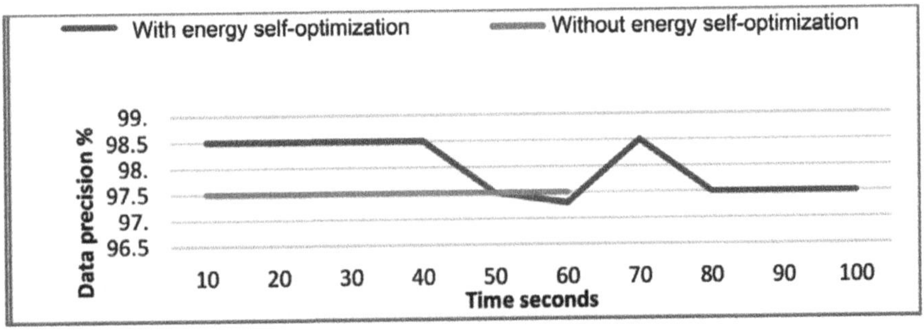

Figure 4.11. *Average precision of the information recorded in scenario 4. For a color version of this figure, see www.iste.co.uk/chalouf/intelligent.zip*

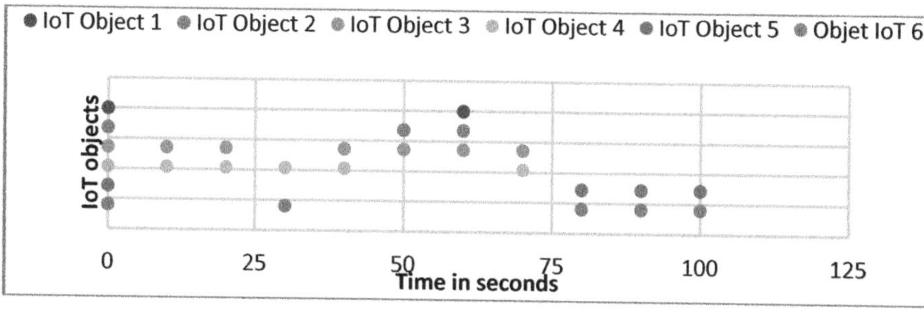

Figure 4.12. *Selection of IoT objects over time in scenario 4. For a color version of this figure, see www.iste.co.uk/chalouf/intelligent.zip*

4.7. Conclusion

Energy self-optimization is needed in an IoT environment to extend the lifespan of the IoT system containing IoT objects with energy constraints. Different techniques and mechanisms proposed for different IT environments have been studied throughout this chapter. To minimize the energy consumption of IoT objects, we have presented several research publications that specify techniques adapted to this domain. Then, we have proposed in this chapter an evaluation system based on a Mamdani type fuzzy logic model as well as an energy self-optimization algorithm. This system and the associated algorithm make it possible to select the objects to activate to offer an IoT service based on several criteria such as distance, the precision and residual energy. We have demonstrated thanks to evaluating performances of several simulation scenarios that this energy self-optimization processes make it possible to extend the lifespan of an IoT system by controlling, intelligently and autonomously, the activation of objects responsible for recording information while still respecting the requirements of corresponding IoT services.

4.8. References

Ashraf, Q., Habaebi, M., Sinniah, G., Ahmed, M., Khan, S., Hameed, S. (2014). Autonomic protocol and architecture for devices in Internet of Things. In *IEEE Innovative Smart Grid Technologies – Asia*. IEEE, Kuala Lumpur.

Baker, A., Lafata, J.E., Ward, R.E., Whitehouse, F., Dinive, G. (2001). A web-based diabetes care management support system. *Joint Commission Journal on Quality Improvement*, 27(4), 179–190.

Claudia, C., Guitierrez, R., Michel, R. (2010). e-Health monitoring applications: What about data quality? In *Health Ambient Information Systems Workshop 2011*. HAMis, Gramado.

Dernoncourt, F. (2011). Introduction à la logique floue [Online]. Available at: https://www.researchgate.net/publication/267041176_Introduction_a_La_Logique_Floue.

Dobson, S., Denazis, S., Fernández, A., Gaïti, D., Gelenbe, E., Massacci, F., Nixon, P., Saffre, F., Schmidt, N., Zambonelli, F. (2006). A survey of autonomic communications. *ACM Transactions on Autonomous and Adaptive Systems*, 1(2), 223–259.

IGI Global (2019). What is network lifetime [Online]. Available at: https://www.igi-global.com/dictionary/network-lifetime/20144 [Accessed 11 June 2019].

ISO (1994). ISO 5725-1:1994, accuracy (trueness and precision) of measurement methods and results – Part 1: General principles and definitions [Online]. Available at: https://www.iso.org/obp/ui/#iso:std:iso:5725:-1:ed-1:v1:en [Accessed 28 September 2019].

ITU-T (2014). Common requirements of the Internet of Things [Online]. Available at: https://www.itu.int/rec/T-REC-Y.2066/en [Accessed 4 February 2019].

Jaxenter (2018). Energy sources and power management in IoT sensors and edge devices [Online]. Available at: https://jaxenter.com/energy-sources-power-management-iot-sensors-edge-devices-145006.html [Accessed 11 June 2019].

Khalil, A., Mbarek, N., Togni, O. (2017). Service level guarantee framework for IoT environments. In *First International Conference on Internet of Things and Machine Learning*. ACM, Liverpool.

Khalil, A., Mbarek, N., Togni, O. (2018). IoT service QoS guarantee using QBAIoT wireless access method. In *4th International Conference on Mobile, Secure, Programmable Networks*. MSPN, Paris.

Kim, J.W., Barrado, J.R.R., Jeon, D.K. (2015). An energy-efficient transmission scheme for real-time data in wireless sensor networks. *Sensors – An Open Access Journal*, 15(5), 11628–11652.

Konakalla, S.V. (2014). A star algorithm. Document, Indiana State University, Bloomington.

Li, J.M., Chen, J.J., Cheng, H.H., Tseng, Y.C. (2013). An energy-efficient sleep scheduling with QoS consideration in 3GPP LTE-advanced networks for Internet of Things. *Emerging and Selected Topics in Circuits and Systems*, 3(1), 13–22.

Li, X., Luo, C., Ji, H., Zhuang, Y., Zhang, H. (2020). Energy consumption optimization for self-powered IoT networks with non-orthogonal multiple access. *International Journal of Communication Systems*, 33(1), 1–15.

Masdari, M. and Naghiloo, F. (2017). Fuzzy logic-based sink selection and load balancing in multi-sink wireless sensor networks. *Wireless Personal Communications*, 97(2), 2713–2739.

Mathworks (2019). Mamdani and Sugeno fuzzy inference systems [Online]. Available at: https://fr.mathworks.com/help/fuzzy/types-of-fuzzy-inference-systems.html;jsessionid=3d8750953818e96ed6ee074c149b [Accessed 19 September 2019].

Rana, K. and Zaveri, M. (2011). A-star algorithm for energy efficient routing in wireless sensor network. *Trends in Network and Communications, Communications in Computer and Information Science*, 197, 232–241.

Salehie, M. and Tahvildari, L. (2005). Autonomic computing: Emerging trends and open problems. In *Workshop on Design and Evolution of Autonomic Application Software*. IEEE, Saint-Louis.

Scribd (2019). Motivations of fuzzy logic [Online]. Available at: https://www.scribd.com/document/172969317/Motivations-of-Fuzzy-Logic-doc [Accessed 4 April 2019].

Singh, A.K., Alkesh, A., Purohit, N. (2011). Minimization of energy consumption of wireless sensor networks using fuzzy logic. In *International Conference on Computational Intelligence and Communication Networks*. IEEE, Gwalior.

Sinreich, D. (2006). An architectural blueprint for autonomic computing. White paper, IBM, Hawthorne.

Weyns, D., Ramachandran, G., Singh, R. (2018). Self-managing Internet of Things. In *SOFSEM 2018: Theory and Practice of Computer Science*. OCG, Krems.

Wyatt, J. and Sullivan, F. (2005). What is health information? Technical Report, BMJ.

Zouinkhi, A., Ayadi, H., Val, T., Boumedyen, B., Abdelkrim, M.N. (2020). Auto-management of energy in IoT Networks. *International Journal of Communication Systems*, 2019, 33(1), 1–16

5

Toward Intelligent Management of Service Quality in the IoT: The Case of a Low Rate WPAN

Guillaume LE GALL[1], Georgios Z. PAPADOPOULOS[1], Mohamed-Aymen CHALOUF[2] and Olivier TOGNI[3]

[1]IRISA, IMT Atlantique, Rennes, France

[2]IRISA, University of Rennes 1, Lannion, France

[3]LIB, Bourgogne Franche-Comté University, Dijon, France

The Internet of Things (IoT) makes it possible to connect devices with reduced processing capacities to the network, using wireless communications links. These technologies are particularly useful for improving the control and monitoring of industrial processes, as they make it possible to deploy sensors and actuators, often battery-powered ones, more easily in a production site. These objects are used to form Low Rate Wireless Personal Area Networks (WPANs), recording only the information necessary for the infrastructure while still being low cost, both with regard to energy and maintenance. However, they cover applications that are sometimes critical, or which are used in cases where an improvement in the network's efficiency means better functioning of the process in which they are used. It is therefore important to work to improve the quality of service in such networks. To do this, there are well-known methods, such as time division and channel hopping, or more formally Time Slotted Channel Hopping (TSCH). It is still possible build upon these techniques, such as with the PAREO method described here. In this chapter, we will first present what the IoT is in sections 5.1 and 5.2, and what the protocols are that are used in this domain. In sections 5.3–5.5, we will then focus on IEEE Std. protocols 802.15.4-2015 TSCH and the Routing Protocol for Low-power and Lossy Networks (RPL), explaining how they can be used, and what their assets

Intelligent Security Management and Control in the IoT,
coordinated by Mohamed-Aymen CHALOUF.

are for increasing the quality of service in IoT networks. Finally, in section 5.6, we will detail our propositions with the aim of obtaining intelligent management of the quality of service in these networks.

5.1. Introduction

On the Internet, the transmission of data, whatever it is (file exchange, multimedia, videoconference, etc.), is based on the Internet Protocol (IP), which makes it possible to have a single network for different uses. The fact that this protocol makes it possible to access the network in a uniform way has facilitated the work of application developers, and has contributed to its success. According to the specification of Internet Protocol version 4 (IPv4), the aim of this protocol is mainly to provide a means of routing packets from the source to the destination, and to allow the fragmentation and reassembly of long packets (Information Science Institute 1981). Also, the IP was designed to perform best effort routing. The drawback it presents is that the routing process it uses is applied homogenously to all types of data. In other words, it offers no guarantee that data will be routed well all the way to their destination, nor on the time that this will take. Fairly quickly, a number of services, such as multimedia and real-time communications, necessitated the implementation of new mechanisms for managing Quality of Service (QoS). These mechanisms have made it possible to implement differentiated policies of QoS, to meet the demands of different applications.

In the literature, a universal definition of QoS has not been found. In fact, each community has its own definition. The ITU-T (International Telecommunications Union – Telecommunication) defines the QoS as a set of criteria for quality required for operating one or more objects. For the Internet Engineering Task Force (IETF), QoS designates the parameters characterizing the demands and constraints of applications as well as those of the whole network. Implementing QoS requires implementation of basic mechanisms that can be regrouped into different plans: data, control and administration. The data plan includes mechanisms at the level of the network devices processing the data. This includes the classification, conditioning, marking, scheduling, queue management and congestion control. The control plan includes mechanisms permitting configuration of resources along the path taken by the data. This includes routing based on information from QoS, admission control based on the state of the resources available, admission control based on a control policy and the QoS signage. The administration plan includes mechanisms for managing and administrating data flows. This includes metrology, contract management, placing QoS parameters and network parameters in correspondence and the QoS policy. There are many QoS parameters making it possible to evaluate the performances of an IP network. Among these parameters, we find the latency (transmission delay), the jitter (delay variation), the bandwidth (or flow rate) and the packet loss rate.

Among the approaches that make it possible to consider the QoS in IP networks, we find the IntServ (Braden *et al.* 1994) which marked the IETF's desire to define an architecture able to support real-time applications based on a resource reservation mechanism, specified by the RSVP protocol (O'Dell *et al.* 1997), in the network and applied to all the routers for a specific flow. Despite its robustness, the IntServ model had a major drawback, and that is the problem of scaling. To overcome this problem, the IETF introduced the DiffServ model (Black *et al.* 1998) which introduced the notion of flow aggregation. Instead of offering a QoS for each flow (as in the IntServ approach), DiffServ offers services by aggregates by pushing flow processing to the ends of the network, which allow the core routers to process the packets according to their class, attributed at the domain boundary. Moreover, there are other traffic engineering mechanisms that make it possible to improve QoS by using the resources available in a network such as Multiprotocol Label Switching (MPLS) (Viswanathan *et al.* 2001) or Generalized Multi-Protocol Label Switching (GMPLS) (Mannie 2004) more efficiently.

For some years, we have been witnessing the emergence of the IoT. This "new" paradigm consists of connecting not only computers but many other devices to the Internet to allow them to transmit and share data collected in the real world. The development of the IoT has benefitted from omnipresent connectivity thanks, especially, to wireless technologies such as WiFi, 4G/5G cellular networks or even low consumption networks – long range (LPWAN: Low Power Wide Area Network) and based on technologies such as LoRa, Sigfox, LTE-M and NB-IoT. The multiplication of IoT devices has drawn out new constraints on networks, especially on QoS. Some applications, particularly industrial ones, could benefit from the IoT, provided that communications will be able to meet some needs in terms of latency or delivery rate. Indeed, just like at the start of the Internet, the IoT has been designed from a best effort angle. In addition, the use of wireless technologies accentuates the difficulty of implementing QoS, because of the problems of interference, collisions and low data rate.

Moreover, the devices, or objects, involved in IoT applications have limited computing and memory capabilities. These objects are often supplied by a battery. A large number of IoT applications are based on the redundancy of objects in a zone. For example, border surveillance or even the detection of forest fires requires the use of a very large number of objects whose lifespan depends only on the battery life. Indeed, the cost of repairing, recharging or changing the battery is higher than the cost of the object itself. Thus, the energy consumption of objects is an additional constraint for a large number of applications, and it becomes vital to optimize object lifespan. As the radio function of communicating objects consumes energy, access to the medium and its management can play a major role in optimizing object lifespan and hence the efficiency of IoT applications. Also, different processing tasks carried out by these objects, especially those linked to data routing when they form a mesh network, can cause a rapid decrease in object lifespan. Thus, optimizing the routing process in IoT networks could contribute to increasing object lifespan.

5.2. Quick overview of the IoT

5.2.1. *The micro-IPv6 stack*

Since its invention in the 1990s, the World Wide Web has relied on open protocols standardized by the Internet Engineering Task Force (IETF), mainly Hypertext Transfer Protocol (HTTP) (Berners-Lee *et al.* 1996; Fielding and Reschke 2014). The fact that this proposes a unified and simple interface allows any machine easy access to resources on the Internet network. HTTP relies on the transport layer Transmission Control Protocol (TCP), which offers fragmentation functions for requests and responses, as well as functions that serve to make communications more reliable and to retransmit lost packets. Finally, on the Internet, exchanges are made on the basis of the IP, whether version 4 or 6 (Information Science Institute 1981; Hinden and Deering 1998; Deering and Hinden 2017).

This protocol stack has shown its efficiency, and it is very widely used for developing applications that use the Internet. However, these protocols have been designed to rely on computers with good calculating capacities, and using high-speed links. On the other hand, they are often too "heavy" for so-called embedded systems, that is, only those with a micro-controller with limited memory, often so-called "objects". Also, the communication links these objects possess most often have a limited flow rate.

So that these objects can benefit from the efficiency of the Web, and to permit them to communicate, it was necessary to define new protocols, adapted to the objects' capacities. This was how the IoT was born (IETF 2020).

Here, the network layer and the way in which it is operated are different from Internet Protocol version 6 (IPv6). However, to keep the flexibility of the Internet network, an adaptation layer is used: IPv6 over Low-Power Wireless Personal Area Networks (6LowPAN) (Montenegro *et al.* 2007). Thus, a node situated on the Internet network will be able to reach another node located in the object network in a transparent way, using IPv6. The role of 6LowPAN is to ensure this adaptation, by compressing the IPv6 headers.

Networks of communicating objects can be organized in complex topologies, which can evolve dynamically. It is therefore called a mesh network. Also, when a packet moves through the network of objects, it should be routed toward its destination. This is the role of the routing protocol, and in particular of RPL (Alexander *et al.* 2012), which also has features that allow the automatic configuration of the network topology depending on the quality of the links and their evolution.

To limit the size of the packets that move through the network of objects, the User Datagram Protocol (UDP) (Postel 1980) is preferable to TCP. As it is very simple, it has the advantage of having a very small header, and its implementation has a "light" memory footprint in the device.

For the application layer, Constrained Application Protocol (CoAP) (Shelby *et al.* 2014) is the counterpart of HTTP for the IoT. Being much lighter, it reuses the powerful concept of Representational State Transfer (REST) Application Programming Interface (API), notably because a client/server relationship exists and the state of the communication is maintained by the client. It also uses codes to indicate the nature of the responses, once again very similar to those of HTTP. The standard that normalizes CoAP specifies that it is used over UDP at the transport layer. As the latter does not benefit from an acknowledgment mechanism or retransmission in case of a loss, CoAP specifies its own mechanisms for making exchanges reliable.

A particularly useful feature of CoAP is Observe. In general, the node in the network of objects will have sensors, and will expose data from these through a CoAP server. The client that will consume these data can make periodic GET requests to follow its evolution. It can also execute a GET request by indicating the Observe option in the header. This indicates to the server that this client wishes to be notified when the value of the resource changes. Using this feature makes it possible to reduce the traffic, and also to be more reactive to changes in the measured values.

An overview of protocol stacks of the classic Web and the IoT is presented in Table 5.1.

Application	HTTP
Transport	TCP
Network	IPv6
Link & Physical	Ethernet, Wi-Fi...

Web protocol stack

Application	CoAP
Transport	UDP
Network	RPL
Adaptation	6LowPAN
Link & Physical	IEEE 802.15.4-TSCH

IoT protocol stack

Table 5.1. *Model Open Systems Interconnection (OSI) in "classic" Internet protocol stacks and the Internet of Things*

Since IoT protocols are compatible with Internet standards, it is possible to interact directly with them from the network. Some applications nevertheless use an application gateway to expose information from objects via the HTTP.

5.2.2. *Technologies for the IoT*

Choosing an IoT communications technology makes sense when a classic communication technology does not meet the constraints. For example WiFi, although it offers a high data rate, consumes too much energy for an object that needs to run on battery power over a period of months, or even years.

Various access technologies exist for the IoT. Even though they generally meet the same needs, they do not all have the same features, nor the same complexity. Some rely on open standards, others are proprietary.

In a wireless network, there is generally at least one node that has the role of concentrator and/or coordinator. It is this node that will act as a gateway between the network of objects and the Internet. It is around this node that the topology of the network will be built.

The simplest topology to set up is the star topology: all the nodes in the network have a single link with the concentrator, and all their communications go through it. This way of organizing the network is valid as long as all the nodes are within range of the concentrator, and their number remains reasonable. A star topology can be very useful and appropriate in a long-range network, for example a LoRa network, or a cellular network, with a base station that serves objects in the surrounding area. But the limitations are soon apparent when a short-range access technology is used, as it offers little flexibility.

When the topology is more complex, and when some nodes are relays of other nodes, it is called a mesh topology. This type of topology makes it possible to have a network of objects that spreads over vast areas, and allows great flexibility. But it also requires implementing a system that will build this topology. It is possible that the routing protocol specifies the means for building and maintaining the topology. This is the case with RPL. An example of a star topology and a meshed topology is presented in the diagram of Figure 5.1.

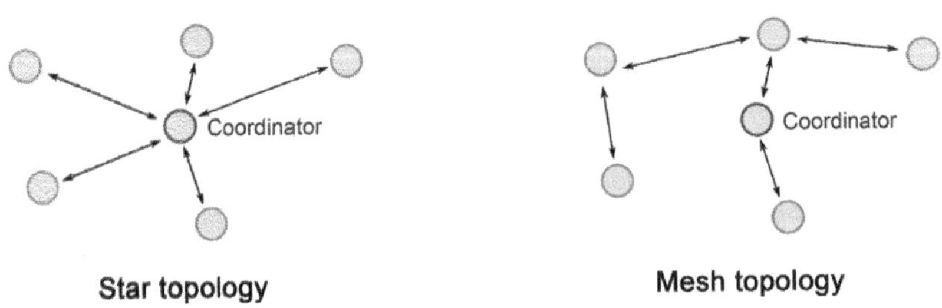

Figure 5.1. *Example of star topology and mesh topology in an IoT network. For a color version of this figure, see www.iste.co.uk/chalouf/intelligent.zip*

The remainder of this section presents a number of examples of access technologies used in IoT networks.

5.2.2.1. *Bluetooth Low Energy*

Bluetooth is a popular choice for short-range communications. Created at the end of the 1990s, it is used in many electronic products endowed with a wireless interface. It is a protocol that specifies the way of using the network from the physical layer to the application layer.

In its Low Energy version, the power used is less than in the classic Bluetooth version, and it is designed as a protocol for sending short frames repeatedly. In version 5 of the standard, the physical layer allows a data rate of as much as 2 Mb per second.

Bluetooth Low Energy (BLE) uses the Industrial, Scientific and Medical (ISM) frequency band between 2,400 and 2,480 MHz, which is divided into channels of 2 MHz width. Therefore, there is no guard frequency left unused between the channels. Three of these channels are reserved for announcing the network.

BLE is simpler than classic Bluetooth, and designed to be used in IoT applications. It has features that are beneficial in terms of QoS, especially channel hopping, which here consists of changing the channel at each transmission for a single link (Yazicigil *et al.* 2018), as well as channel blacklisting, which consists of temporarily stopping to use the channels for which the quality is too low (Bluetooth SIG 2016).

Although BLE allows interoperability with classic IPv6 networks, and makes it possible to create mesh networks, these features are more recent and perhaps less advanced than in the IEEE 802.15.4 and RPL stack. Indeed, the BLE Mesh standard dating from 2017 specifies that the network is a "managed-flood-based mesh network". In other words, this mesh network does not have an efficient routing mechanism; the nodes simply forward the packets, but still have some mechanisms to avoid saturating the network with packets that would be stuck in a loop. The standard specifies that a routing mechanism could be specified in a later version (Bluetooth Mesh Working Group 2017).

5.2.2.2. *WiFi HaLow*

Normalized by IEEE 802.11ah, WiFi Halow is a protocol that was published in 2017. This version, which uses the 900 MHz band that does not require a license, was designed to have lower energy consumption than "classic" WiFi which consumed too much energy for IoT applications. It also aims to have more extensive coverage than the latter. It is in fact an adaptation of IEEE 802.11ac, with a modified physical and link layer (Meera and Rao 2017; Qiao *et al.* 2018).

WiFi Halow was designed for point-to-point communications, or star topology networks. One of its main advantages is that it natively supports IPv6. One of the applications possible with this technology is linking a IEEE 802.15.4 network to the Internet network (Meera and Rao 2017; Qiao *et al.* 2018).

For this technology, access to medium is achieved by Enhanced Distributed Channel Access (EDCA), a derivative of Carrier Sense Multiple Access with Collision Avoidance (CSMA-CA). This protocol, which is the more often used to collect periodic information of small size, has a feature making it possible to pick up report alert messages quickly, called Restricted Access Window (Khorov *et al.* 2019).

5.2.2.3. *WirelessHART*

WirelessHART is a wireless communication protocol for industrial networks, introduced in 2007 and normalized by IEC 62591 since 2009, of which the latest version was published in 2016. It was designed as a wireless evolution of the Highway Addressable Remote Transducer (HART) protocol.

The HART protocol is called a hybrid protocol, which uses digital communications over an industrial process control link using an analog current signal. This protocol, which has existed since the 1980s, also standardizes an API that defines how to communicate with a sensor or actuator that uses HART to communicate (FieldComm Group 2020).

WirelessHART uses IEEE Std. 802.15.4-2006 for physical and link layers in the ISM 2.4 GHz band. It uses both an access mode Time-Division Multiple Access (TDMA) and a channel hopping mechanism, that is, channel hopping at each transmission. The duration of a timeslot, interval of time dedicated to a transmission, is fixed by the standard at 10 ms. The network manager, an entity located outside of it, is responsible for building a transmission schedule, and communicates the parts of this schedule to the nodes involved. It is possible to use blacklisting to prevent the use of lower quality channels. These channels are configured by the system administrator (Nixon and Rock 2012). It should be noted that the standard does not specify which is the channel hopping sequence (Petersen and Carlsen 2011).

At the network layer, a mesh topology is used, in which all the nodes that participate in the network have the capacity to route packets. The technique used to route the packets is Graph Routing, which makes it possible to add redundancy in the routes (Shi *et al.* 2019).

The transport layer defined in WirelessHART has (optional) acknowledgment and fragmentation features (Nixon and Rock 2012). The application layer is HART.

5.2.2.4. *ISA100*

ISA100, or more precisely ISA100.11a, is a protocol developed by the International Society of Automation, a consortium based in the United States.

Published for the first time in 2009, it is a protocol aimed mainly at industrial networks and applications. The standard has already been available in IEC 62734 since 2014. It is fairly similar to and often compared to WirelessHART (Nixon and Rock 2012).

Like WirelessHART, ISA100 uses IEEE Std. 802.15.4-2006 for the physical and link layers in the ISM 2.4 GHz band, except that it proposes a slightly modified version of the Medium Access Control (MAC) layer making it incompatible with the original version of IEEE 802.15.4. The link layer also uses TDMA and channel hopping, but with more configuration options. The duration of a timeslot is fixed by default to 10 ms, but is configurable. As for channel hopping, several possibilities exist. First of all, there is slotted hopping, where devices change channel at each timeslot, but also slow hopping, where the devices use a single channel during a period ranging from 100 to 400 ms. Slow hopping is generally used for a contention-based period, that is, where devices attempt to access the medium concurrently. Finally, it is possible to use hybrid hopping, which is an alternance between slotted hopping and slow hopping (Rezha and Shin 2014). ISA100 defines five predefined sequences for channel hopping, from among which the network administrator can choose (Petersen and Carlsen 2011).

Not all ISA100 equipment necessarily has the capacity to announce the network, or to route packets from other nodes. This peculiarity not only leads to great flexibility, but also requires more planning before the nodes are deployed. The network stack defined in the standard is based on 6LowPAN. Routing happens using mesh in subnets of the wireless network at the link layer, and by using classic IPv6 routing outside of them. The transport layer is UDP, with an enhanced integrity control. One of the main differences between ISA100 and WirelessHART is that the standard specifies completely how the backbone of the network should be implemented (backbone) from the production site (Nixon and Rock 2012).

The application layer uses two main mechanisms: publication and warnings. The standard also proposes extensions that enable each vendor to enrich their communications and to adapt to specific needs (Lydon 2011). The protocol also proposes a tunnel feature, which allows for exchanging packets from any application layer on top of ISA100. For example, it is possible to use HART above ISA100 (Nixon and Rock 2012).

The criticism often made of ISA100 is that the broad flexibility it offers means that objects by two different constructors will not necessarily be interoperable, even if they are both compatible with the standard (Petersen and Carlsen 2011).

5.2.2.5. *IEEE 802.15.4 standard*

IEEE Std. 802.15.4-2015 defines a communication protocol for low data rate, low energy consumption and short-range wireless devices (IEEE SA 2016). The first version of the standard, which appeared in 2003, focused on a means of accessing the medium called Carrier Sense Multiple Access (CSMA), which consists of listening to the state of the medium before transmitting, and waiting for a random duration if it is busy. When this technique is used, a single channel is chosen, and all the nodes in the network use and share it. In the 2015 version, another technique for accessing the medium made its appearance: TSCH. It is detailed in section 5.3.

The standard proposes using several frequency bands, ranging from 169 to 2.4 GHz. A frequency band is designated by its central frequency. Each band is divided into channels, and it is possible to find the central frequency of each channel with the help of a formula. For example, for the 2.4 GHz band, the frequency of each channel is obtained by :

$$F_c = 2\,405 + 5\,(k - 11)\; MHz, \;\; k \in [11; 26] \qquad [5.1]$$

In the 2.4 GHz band, the channels are therefore distributed every 5 MHz. The bandwidth used for communications is in fact 2 MHz, and the rest of the space around the channel, called guard frequency, is left unused to avoid any risk of overlap between the channels.

When starting the network, the coordinator will announce it by periodically transmitting frames called beacons. These frames contain information on the network and how to join it, and mainly its identifier. A node that has succeeded in joining the wireless network can also transmit beacon frames in its turn, to allow other devices to join it. A device wishing to join the network of objects will start to scan all the channels. As soon as it has received a beacon, it can start the association procedure (IEEE SA 2016).

5.2.3. *IoT and quality of service*

Communication technologies for the IoT are numerous. They have been defined to allow communication between objects or even with the Internet. These technologies are intended to be adapted to the needs of the IoT, especially with regard to objects' often very limited capacities, in terms of CPU, memory or even energy. A choice that seems clear for interconnecting these objects is to use a simple access technology, for example based on CSMA, and to choose a single physical channel for all communications. Such a network is then subject to losses, and adopts a best effort operating mode. In other cases IoT technologies specify a complete and fairly strict protocol stack, going up to the application layer, which covers some use cases but does not provide the network of objects with the flexibility of the protocols of the Internet.

Some access technologies, such as those described above, implement additional mechanisms to improve network performances such as, among others, new techniques for accessing the medium, frequency hopping and routing protocols. Certainly, these technologies and these efforts have enabled the development of many IoT applications. Nevertheless, for some other applications, a higher level of QoS is needed. To meet the needs of these applications in terms of QoS, changes in these technologies have been proposed, such as IEEE Std. 802.15.4-2015 TSCH, which is based on dividing time into cells and on using multiple channels in parallel within a single wireless network, making it possible to increase the bandwidth available in the network. In this chapter, we propose using this technology to ensure communications in an IoT network. To have good performances, a network based on IEEE 802.15.4 TSCH needs an efficient scheduling mechanism. In addition, the choice of routing protocol can both determine and have consequences on network performance. In what follows, we introduce our approach based on a judicious choice of scheduling mechanism, linked to use of a multipath RPL routing, to move toward smart management of QoS in IoT networks.

5.3. IEEE 802.15.4 TSCH approach

In the 2015 version of the IEEE Std. 802.15.4, even though the CSMA technique is still present, another technique has been introduced: TSCH.

This relies on two principles: time division into cells and the use of multiple channels for a single wireless network.

In this approach, the aim is to reduce the risks of collisions between packets by sharing the time and the channels, and by attributing spaces in advance to transmit to the nodes, called timeslots. According to the standard, a timeslot is a short interval of time that should be long enough to allow two nodes to exchange a frame and an acknowledgment. It is also an opportunity for a node to transmit or receive a packet. To organize communications, the nodes use a slotframe, which is a pattern containing the communication scheduling information, and is repeated over time. Each timeslot in the slotframe can either be shared or dedicated. If the timeslot is shared, all the nodes can use it, and potential collisions are solved using the CSMA method. If the timeslot is dedicated, the potential transmitter and receiver of the transmission are pre-established, and no other node should transmit in this timeslot. One example of a slotframe is shown in Figure 5.2. It should be noted that the size of the slotframe is configurable and depends on needs in terms of network use. It is common to have a slotframe length of around 100 timeslots. These are generally configured to have a duration of 10 ms.

As all the nodes in the network act in a synchronized way, they hop from one channel to another in line with transmissions, to minimize the effect of interference on a timeslot in particular. The aim here is to avoid an instance where the same timeslot would always use a single, poor-quality channel in all the slotframes. This technique is called channel hopping.

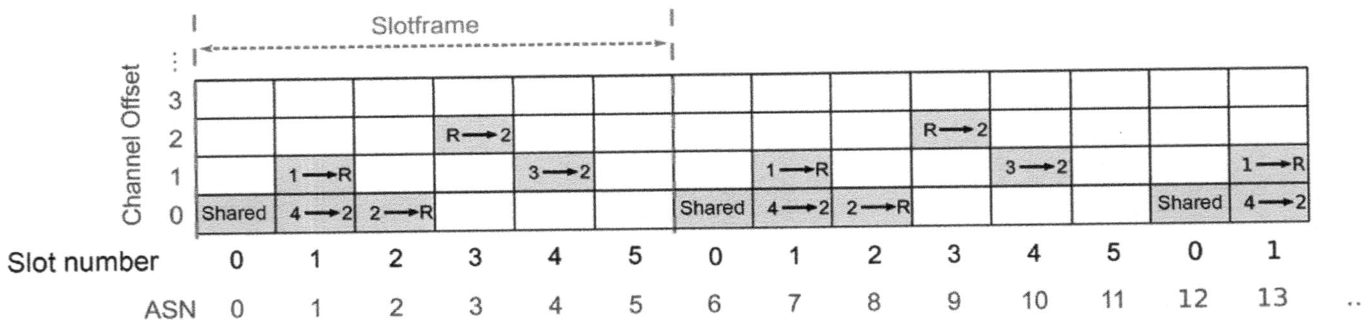

Figure 5.2. *Example of a size 6 slotframe, for a 4-node network and a coordinator using TSCH. For a color version of this figure, see www.iste.co.uk/chalouf/intelligent.zip*

To achieve this, the nodes count the number of timeslots which have passed since network formation. This variable is called Absolute Slot Number (ASN), and is also announced in the beacons which announce the network. When a node must transmit and receive in a timeslot, it finds the physical channel it should use from the channel offset and the ASN, and using the formula [5.2]:

$$CH = HSL((ASN + ChannelOffset)\%NumChannels) \qquad [5.2]$$

In this equation, CH is the physical channel that will be used for the transmission, the HSL function, which means Hopping Sequence List, is in fact a permutation of channel identifiers, and is known in advance by all the nodes, or announced in a beacon. Thus, it is possible for any node that has joined the network to make a link between a channel offset, a *slot* number and a physical channel that will be used to transmit or receive.

It should be noted that the standard proposes a default channel hopping sequence, and it also offers the possibility to use a custom sequence. If a custom sequence is used, it must be loaded in the nodes in advance, or it is given in the beacons (IEEE SA 2016).

What is not specified in the standard, and is nevertheless essential to the proper functioning of the network that uses TSCH, is the transmission scheduling, or schedule. According to the standard, the means of constructing and distributing the information it contains is left to an "upper layer" that will therefore implement a scheduling technique scheduling.

5.4. Transmission scheduling

5.4.1. *General considerations*

The method of scheduling transmissions in a network that uses TSCH depends on the applications that will use this network. In scientific literature, many publications on the subject can be found (Hermeto *et al.* 2017). There are two main families of techniques making it possible to construct the schedule: centralized techniques, where a central controller has an overall vision of the network and decides for each of the nodes, and distributed techniques, where the nodes negotiate locally the dedicated timeslots they will use.

Several criteria are taken into account for choosing the schedule. As stated above, the nodes that participate in IoT networks are often battery powered. Also, with TSCH, the nodes only need to turn on their radio during timeslots during which they are likely to receive a packet, or when they have a packet to transmit. This will allow them to make substantial energy saving compared, for example, to a technique such as CSMA, where the nodes are in competition to access the medium, and are likely to receive a packet at any moment.

Moreover, the amount of data sent by the nodes is very important in deciding the schedule. If a node sends few packets, it is desirable not to program a large number of transmissions for it in the slotframe. Here, we also note that a node that needs to route many packets from its neighbors will consume more.

The quality of the link should be taken into account when building the schedule. In fact, if one link is of low quality compared to the others, more packets will be lost. To improve the QoS locally and to avoid congestion phenomena, it is therefore desirable to give the nodes that use this link opportunities for retransmission.

Another criteria to take into account is topology. If it changes, some timeslots will become unusable, while others will need to be allocated. We see a strong coupling between topology and scheduling.

Finally, it is possible that the needs of applications as well as the environment vary over the lifespan of the wireless network. One example is the malfunctioning of devices that seek to send many warning messages, or other uses of the 2.4 GHz band which would increase their flow, thus generating more interferences on the channels that overlap with the IEEE 802.15.4 network. This consideration becomes even more important when the nodes that participate in the network of objects move physically in the environment.

To return to choosing a centralized approach rather than a distributed approach, a centralized approach relies on the fact that a single central controller makes all the choices, most often with a maximum of information on the qualities of the links, or in any case an overall vision of the network. On the other hand, this strategy involves increasing the traffic related to slotframe management in the network, as all this information should pass via the coordinator. This concern is even truer for large networks. With a distributed strategy, decisions are taken locally, and on the basis of partial information. However, this strategy has the advantage of needing only a small amount of communications, since the exchanges for building the schedule are only generally done with neighboring nodes. It is therefore better adapted to large networks.

5.4.2. *Scheduling in the literature*

In this section, examples of scheduling strategies for networks using TSCH are presented. These strategies are varied, there is none that is strictly better than the others, and it is therefore necessary to choose such a strategy depending on the network and the applications.

5.4.2.1. *TASA*

In Palattella *et al.* (2012), the authors propose their Traffic Aware Scheduling Algorithm (TASA). This is a centralized scheduling strategy, which uses the topology and traffic to build the slotframe, by means of methods taken from graph theory.

According to the authors, this strategy is particularly efficient in cases where the traffic patterns are relatively constant over time. Also, they make the hypothesis that the traffic is solely uplink, that is, in the direction of the network coordinator. To build the slotframe, they assume that the coordinator knows the topology and uplink rate at each node, and using the vertex coloring and vertex matching methods.

5.4.2.2. *DeTAS*

Decentralized Traffic Aware Scheduling (DeTAS), presented in Accettura *et al.* (2013), is closer to a distributed solution, as some of the processing is carried out at the nodes. Here, the authors again take up the hypothesis of an upstream traffic that is relatively constant over time.

DeTAS uses the information coming from RPL to build a slotframe. Each sub-tree that departs from the root node is called a routing graph. DeTAS first builds a scheduling for each of these sub-trees called micro schedules and then aggregates them into a macro schedule. Also, the micro schedules are built such that only even rank nodes will transmit in the timeslot pairs, if the even-schedule is used, with the aim of reducing the risk of collision. If the odd-schedule is used, it is the opposite. One important point of this technique is that each node will count the number of packets it transmits per slotframe, including the packets that it routes for its neighbors, to know how many dedicated upstream timeslots should be allocated to it.

5.4.2.3. *Orchestra*

As for Orchestra it is an entirely distributed strategy, introduced in Duquennoy *et al.* (2015). Unlike the previous approaches, it addresses cases where the topology and traffic schemas are unstable. According to the authors, the aim of this strategy is to "build a non-deterministic network while still exploiting the robustness of TSCH", and requires low energy consumption. One particularity of this algorithm is that it takes advantage of the opportunity to have several slotframes simultaneously, with which the nodes participate at the same time, and which have different priority levels. Thus, each "traffic plane" (application, RPL, TSCH) has its own schedule, the ones associated with the lowest layer having the highest priority.

In Orchestra, each node maintains its representation of the schedule locally, and uses topology information coming from RPL to allocate timeslots. Concretely, when starting the network or when a node joins it, it has a shared timeslot common to all the nodes, in which these are in competition to access the medium. Once the association process is carried out, the nodes use one method (among three) to determine the timeslots they will wish to use. They can have either one reception timeslot over which the neighbors that want to transmit to it are in competition, or one transmission timeslot per node to which all the neighbor nodes should listen, or one dedicated transmission timeslot per node and per neighbor (this is only possible if the slotframe is long enough). One particularity of this algorithm is that the timeslots are not negotiated. Because the attribution method is known to all nodes from the outset, the location of the timeslots in the schedule is deduced by the nodes from the topology RPL, and either one operation on a hash MAC address or based on a unique identifier specific to each node.

5.4.2.4. *LOST*

LOcalized Scheduling algorithm for TSCH networks (LOST) is another algorithm for creating a schedule for TSCH (Zorbas *et al.* 2018). Like Orchestra, it is a completely distributed strategy, where all decisions are taken locally. To create the schedule, it uses the traffic at the level of each node, based on the size of their queues, and also carries out over-provisioning, that is, it allocates more timeslots than needed, making them available to be used for retransmissions. Its particularity is that it proposes a feature for local blacklisting of channels.

Concretely, in the network, the topology is allocated by the RPL protocol, and LOST uses it to know how to build the schedule. The construction of the schedule in LOST happens iteratively, each iteration carrying three stages. First, each node communicates to its neighbors the quantity of packets it needs to transmit and a priority that is calculated on the basis of this number of packets and the RPL rank of the node. Then, each node decides whether it is going to be a requester over the course of this iteration, which is only possible if it has a priority higher than that of its parent. Otherwise, it will be a respondent. Finally, the final stage of the iteration for respondents consists of classifying the requests they have received, to attribute timeslots for transmissions from requesters, and to communicate these to them. An improvement proposed by the authors of this article is local blacklisting of the worst channels, for which they propose a method for choosing the set of physical channels that will be used on a link.

5.5. Routing and RPL

5.5.1. *Routing*

From section 5.2.2 we recall that in a network of objects, the topology can be in star or mesh form. If it is in star form (i.e. if each node has a unique link with the coordinator), there is only a single chance to route the packets: the nodes will by necessity use the only link they have: the link with the concentrator. And if the concentrator must deliver a packet to a node in the network, it will choose the direct link that it has with it.

However, if the network is mesh, then there are many ways to route the packets, and it is necessary to have a mechanism not only to build routes, but also to route the packets.

In particular, in the decision-making process, the routes that have better quality links will be favored, to limit packet loss and retransmissions, which are costly in terms of energy (especially if the nodes are supplied by a battery). It is also desirable to seek to use shortest paths, for the same reasons. Finally, an example where some nodes in the network are battery-powered or others are on mains may also occur.

It is to meet these needs in an industrial context that the Routing Protocol for Low-power and Lossy Networks (RPL) was designed.

5.5.2. *RPL*

As indicated in "Request for Comments (RFC) 5867", which proceeds the creation of RPL, the aim of this protocol is to ensure routing in Low-power and Lossy Networks (LLN), for industrial, commercial, domestic and urban applications (Martocci *et al.* 2010). In this type of application, the nodes, and also the links that connect them to one another, are constrained. These links present limited data rates, sometimes high packet losses, or they are unstable. In addition, in this type of network, the number of nodes can sometimes be very high, as much as several thousand (Alexander *et al.* 2012). RPL supports three traffic schemas: point-to-point, where a node communicates with another (potentially outside the Low-power and Lossy Networks (LLN)), multipoint-to-point, where the network nodes upload information to the coordinator, and point-to-multipoint, which with the help of which the coordinator can diffuse a message to all the nodes in the network. One of the strengths of RPL is that it requires few node configurations upstream, and it includes the mechanisms needed to propagate the network's configuration information.

A network can implement one or more instances of RPL. One instance is characterized by an identifier and an objective function. This latter specifies how the nodes will select their parents to build the topology. Each RPL instance will build one or more Destination-Oriented Directed Acyclic Graph (DODAG), which as its name indicates is oriented toward a root node. Often, it is the network coordinator that will also play the role of gateway to the Internet. The main objective of DODAG is to avoid the loops when routing packets. In this graph, the nodes have a rank, the root node having the smallest. When a node joins the network, it will choose a parent with the help of the objective function, and will then obtain a rank higher than that of its parent in the DODAG. Since the quality of links may vary, RPL proposes mechanisms that make it possible to reconstruct the DODAG. Therefore, in RPL, a topology is defined by a RPL instance identifier, a DODAG identifier, and a DODAG version.

In a DODAG, the root node announces itself to the others with the help of a message called DODAG Information Object (DIO), which will then be propagated along the graph. One node can also solicit its parent at any moment to send it a DIO, with the help of a message called DODAG Informational Solicitation (DIS). These messages make it possible to build upstream routes toward the root in the network. As for the descending routes they are constructed with the help of Destination Advertisement Object (DAO), which the nodes use to announce themselves to their parent, which will then propagate in the direction of the root. The nodes playing the role of parents to other nodes have a feature for aggregating the DAO, to reduce the bandwidth used for signaling in the network.

To adapt to the capacities of the nodes that form the network, RPL proposes several modes of operation: no downward route, where the application traffic can only be routed toward the root; non-storing mode, where all the packets transit through the root, including those exchanged between two nodes participating in the network; storing mode without multicast support, where the packets are relayed in the direction of the root until they reach a common ancestor, then toward the destination node; and finally storing mode with multicast support, which is equivalent to the previous mode except that it allows for sending multicast packets.

RPL has several other features which contribute to improving the QoS. Among these, we find DODAG repair, which allows for rebuilding the topology either globally, or locally, always with the aim of avoiding routing loops. Also, RPL implements mechanisms for detecting these loops, and triggering repairs when this is needed. Moreover, it also limits the operations that a node can carry out when it has just joined the network, thus limiting the risks of instability in the topology. Finally, RPL has just as many error management and traceability mechanisms, making it possible to study the history of the states through which the network has passed if there are problems.

5.5.3. *Multipath*

We anticipate that wireless sensor networks (WSNs) will play an important role in leading industrial processes toward an additional degree of automation in industry 4.0. Indeed, this latter aims to make it possible to monitor and control as many parts of the production chain as possible, while still being able to make changes to it without prohibitive cost. With the WSNs, it then becomes easier to reconfigure the sensors so that they adapt to a new configuration, because they communicate wirelessly and because they do not often require an external power supply.

The problem lies in applying these wireless technologies to the needs of production chains, given that industry requires solutions offering a high level of QoS, while wireless technology generally offers a best effort solution (Frotzscher *et al.*

2014; Kumar S. *et al.* 2014). Thus, the main challenge is the reliability and predictability that the networks should be able to offer to production chains. Normalization processes have been undertaken with the aim of providing deterministic networks, such as those of the *Deterministic Networking* (DetNet) Working Group (WG) at the IETF, but at present, these efforts have been centered mainly on wired networks (Grossman 2019).

Currently, to compensate for packet loss, nodes carry out retransmissions, which impact the jitter, and eventually make the network less predictable. In addition, retransmissions have no effect when faced with situations where a node that should route packets is temporarily inaccessible. To overcome these problems, we work with and above the protocol standardized by IEEE Std. 802.15.4-2015 TSCH (IEEE SA 2016), since it offers features to approach determinism in the transmission of packets beyond a wireless medium.

It is to move further in this direction that a new approach, centered on wireless, is considered at the IETF, and is entitled Reliable and Available Wireless (RAW) (Thubert and Papadopoulos 2019). This is an approach that uses layer 3 with a view to providing a deterministic network, managed intelligently, and which will function on a vast range of radio technologies using a transmission schedule (Thubert *et al.* 2019), which include among others TSCH, 5G and WiFi 6. RAW considers the routing operations along a complex path, called track, with the possibility of having links and paths available in parallel, and also the heterogenous technologies that form these paths. The track highlights many possibilities of redundancy from which RAW tries to benefit to counter the random phenomena inherent to each wireless link. Overhearing, specific to radio technologies, is an example of this.

5.6. Combined approach based on 802.15.4 TSCH and multipath RPL

The IETF has defined a set of functions, called PAREO functions, to enable the multipath and improve the reliability and availability in LLNs.

Packet Automatic Repeat reQuest (ARQ), Replication and Elimination (RE), and Overhearing (PAREO) methodology consists of a set of functions whose aim is to improve reliability, predictability and energy consumption in LLNs. These individual functions, illustrated in Figure 5.3, are complementary from the point of view of this objective, even if they operate relatively independently of one another. The only aspect on which they need to be directly coordinated is the definition of the TSCH schedule, where depending on the functions that are used, the schedule needs to be modified accordingly.

In the following sections, we present each of the functions that form the PAREO method and the way in which they interact.

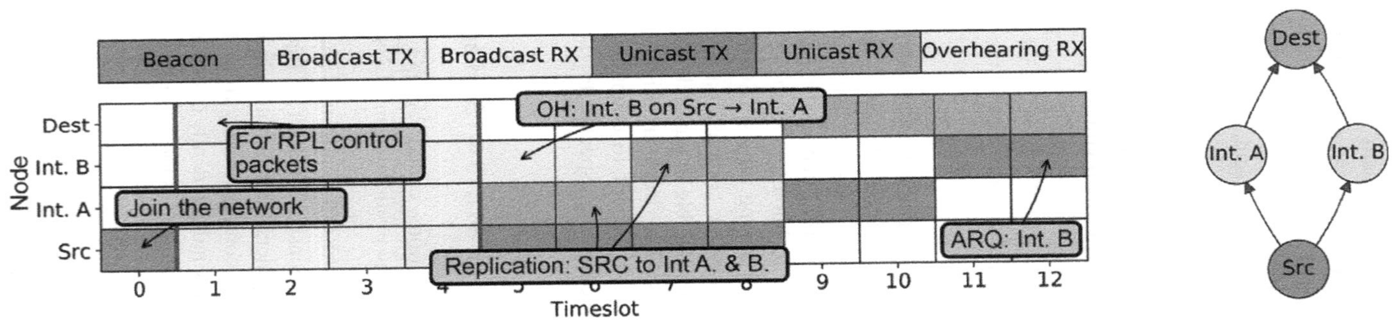

Figure 5.3. *Example of TSCH schedule and topology showing Replication, Overhearing and ARQ operations. For a color version of this figure, see www.iste.co.uk/chalouf/intelligent.zip*

5.6.1. Automatic Repeat reQuest

The ARQ function carries out retransmission of data packets when a previous transmission has failed (Fairhurst and Wood 2002). In our case, we use Automatic Repeat reQuest (ARQ) at the link layer so that the decision to retransmit is taken directly by the transmitter node. To use this function, it is necessary to use an acknowledgment (ACK) control packet for each data packet transmission. Fortunately, TSCH allows cells to be configured to require that the receiver sends an ACK to the transmitter in the same cell, at each data packet reception.

Because in our case, a single packet will be "in the air" at a precise moment and in one cell in particular, it is possible for us to use the Stop-and-Wait ARQ variation, as described below. This function begins by sending a packet configuring a short interval of time to await an ACK (within the same cell). If the ACK coming from the receiver is received, the transmission is marked as successful and finished. If no ACK is received, then the transmission is marked as having failed temporarily, and re-scheduled for sending later. For each data packet, a counter is incremented at each (re)transmission that fails. If the counter reaches a threshold, called the Re-Transmission (RTX) count, then the packet is marked as having failed permanently and is no longer programmed to be sent.

Planning for retransmissions is directly linked to the structure of the TSCH schedule. Depending on this schedule, the delay between a failure and a retransmission attempt may vary significantly: retransmission can take place after a temporary failure, or indeed quite a long time afterward, during the following slotframe.

Even if it is possible to include almost any number of retransmission attempts for each pair of node in a single slotframe, the more cells are reserved for potential retransmissions, the longer the slotframe will be.

5.6.2. Replication and Elimination

The Replication and Elimination (RE) function modifies the routing of packets so that these are sent not only to the Preferred Parent (PP) of each node, but also to the other nodes of the Parent Set (PS) (Papadopoulos *et al.* 2020). In the default implementation of RPL, each node transmits its packets uniquely to one other, the PP.

The problem with this approach is that if the PP starts to fail, all the packets coming from its children will be lost, and this will continue until another parent is selected. However, selecting a new parent involves using one of the repair processes, local or global, from RPL, and these take time, while in these moments, connectivity is unavailable. If a node's queue becomes full while the connectivity is still not re-established, the packets will be lost. With RE, global or local repairs to select a new PP are not avoided, but through having access to two parents or more at the moment a packet is sent, it is possible to maintain connectivity and to avoid losses that would result from a parent being inaccessible.

To permit this, the packet is cloned and each copy is routed toward a different parent, called Alternative Parent (AP) in the PS. Carrying out such packet replication operations creates multiple copies of the same packet, which circulate in the network independently. To avoid a situation where each packet received is a copy with a content also present in a set of packets still in the process of being routed, resulting in a flood, it is necessary to implement an elimination function.

The process of packet elimination uses a unique 32 bit ID added in the header of each packet sent, to identify the duplicated packets. This identifier is placed in an *Option* which extends the IPv6 RPL header in our implementation.

The unique identifier of the last k packets to be routed by each node is saved in a Least Recently Used (LRU) or Last In, First Out (LIFO) type unique-id-history cache. This makes it possible to verify the presence of this ID in the unique-id-history cache when a new packet arrives to be routed. If the ID is present, then the packet is ignored, since this means that a copy has already been routed previously by this node. Otherwise, the unique identifier is added to the cache, which can have the effect of removing another entry if the cache is full. The size k of the unique-id-history cache should be higher than or equal to the largest number of packets that can transit by a node during a time period equal to the greatest end-to-end delay possible between any source and destination in the network.

Selecting an AP is not simple; this is why we have previously proposed a methodology called Common Ancestor (CA) (Koutsiamanis *et al.* 2018, 2019). Its principle is as follows. When a node n chooses an AP, it verifies that the AP selected has in its PS the node that is the PP of n. To enable this selection mechanism, additional information is added in the DIO control messages that the nodes transmit broadcast. This information is the list of IPv6 addresses corresponding to the m lowest-ranking parents (and so those that are preferable) in all node n's parent set.

We have implemented this feature in the operating system that we use, Contiki OS, so that this information is included in DIO messages, in the Directed Acyclic Graph (DAG) Metric Container (MC)[1].

The interaction of Replication and Elimination (RE) with TSCH scheduling here only concerns the need for the existence of cell(s) for the transmission of packets from a node no longer only to its PP, but also to its AP.

1 https://github.com/ariskou/contiki/tree/draft-ietf-roll-nsa-extension.

5.6.3. Overhearing

The Overhearing (OH) function (Lee and Huh 2009) can be used to take advantage of the shared nature of the wireless medium, with the aim of increasing its reliability. This function allows a node to route a single packet to multiple parents simultaneously, which can be supposed to be within range by virtue of their presence in the PS. OH only has a purpose if RE is also used, working in complementary fashion to improve reliability without increasing latency in transmissions.

Since with this function the sender effectively broadcasts the packet the receiver should disactivate its MAC layer filter, two problems may then arise.

First, OH is carried out on the packets that are normally sent using unicast transmissions with acknowledgment. Indeed, several parents may receive the transmission and attempt to answer it with an ACK, which could give rise to a collision between acknowledgments. The consequence would then be that the transmitter would consider its transmission to have failed.

It is therefore important that a single node, the one indicated as receiver in the destination address field, answers with an ACK, whereas all the others receive the packet silently. This capacity is not present by default in TSCH, nor implemented in Contiki OS. We have provided a solution by introducing Overhearing (OH) cells, which accept packets sent to any destination (by disabling MAC filtering), but where the response by an ACK is disactivated, even when this is requested.

Second, it is desirable to control the replication of packets using OH so that only the AP routes the packet, and not any node that receives it via OH.

What is described here is not supported by default in TSCH or RPL. We have therefore considered two ways to implement OH. The first consists of using the TSCH schedule by configuring only the AP that will carry out OH when the node transmits. The second consists of storing the AP address in the packet, so that even if another node receives it, it can compare the address of the AP with its own, and if they are different, they will not route the packet. In our implementation, we have opted for this latter option, as it simplifies management of the TSCH schedule.

The interaction between OH and TSCH scheduling is in fact limited to the existence of reception cell(s), necessary for the OH of packets from one node to its AP.

5.7. Conclusion

Connecting on-board devices that have reduced capabilities and only a single wireless interface to the Internet is a complex problem, especially if we wish to preserve the flexibility of the protocols based on IP. The first IoT applications were designed to be best effort, which may be sufficient in some cases. However, other applications require a minimum level of QoS, in particular in the case of deployment in an industrial context. Many access technologies are available to implement a network of objects. As they all present their disadvantages, it is necessary to choose them well, depending on the applications that use the network.

To overcome these difficulties, recent standardization initiatives, notably from the IETF, have made it possible to improve the QoS in these networks significantly. They rely on a technology that makes it possible to optimize access to the medium and which has been tested: IEEE 802.15.4 TSCH. RPL is used on top of this. And more recent propositions, such as the multipath technique described in section 5.5.3, used in conjunction with the PAREO functions, enable increased network reliability by maintaining reasonable energy consumption in the various nodes that form the network.

The IoT is adapting to meet the needs of applications, and consequent progress in improving reliability has been made in recent years. However, in these networks, there is always a tight coupling between the data that transit at the application layer and the way these links are used, in particular in the case of specific applications from which stem traffic patterns which are not often addressed in the literature. Possibilities remain therefore to improve QoS in these networks.

5.8. References

Accettura, N., Palattella, M.R., Boggia, G., Grieco, L.A., Dohler, M. (2013). Detas: A decentralized traffic aware scheduling technique enabling IoT-compliant multi-hop low-power and lossy networks. In *2nd IEEE WoWMoM Workshop on the Internet of Things: Smart Objects and Services*. IEEE, Madrid.

Alexander, R., Brandt, A., Vasseur, J., Hui, J., Pister, K., Thubert, P., Levis, P., Struik, R., Kelsey, R., Winter, T. (2012). RPL: IPv6 routing protocol for low-power and lossy networks. Internet Engineering Task Force [Online]. Available at: https://rfc-editor.org/rfc/rfc6550.txt.

Berners-Lee, T., Fielding, R.T., Nielsen, H. (1996). Hypertext Transfer Protocol – HTTP/1.0 [Online]. Available at: https://rfc-editor.org/rfc/rfc1945.txt.

Black, D.L., Wang, Z., Carlson, M.A., Weiss, W., Davies, E.B., Blake, S.L. (1998). An architecture for differentiated services [Online]. Available at: https://rfc-editor.org/rfc/rfc2475.txt.

Bluetooth Mesh Working Group (2017). Bluetooth specification – Bluetooth mesh profile. Report, Bluetooth SIG [Online]. Available at: https://www.bluetooth.com/specifications/mesh-specifications.

Bluetooth SIG (2016). Bluetooth core specification v5.0. Report, Bluetooth SIG [Online]. Available at: https://www.bluetooth.com/specifications/blue tooth-core-specification.

Braden, R.T., Clark, D.D.D., Shenker, S. (1994). Integrated services in the internet architecture: An overview [Online]. Available at: https://rfc-editor. org/rfc/rfc1633.txt.

Deering, D.S.E. and Hinden, B. (2017). Internet Protocol, version 6 (IPv6) specification [Online]. Available at: https://rfc-editor.org/rfc/rfc8200.txt.

Duquennoy, S., Al Nahas, B., Landsiedel, O., Watteyne, T. (2015). Orchestra: Robust mesh networks through autonomously scheduled TSCH. In *13th ACM Conference on Embedded Networked Sensor Systems*. ACM, Seoul.

Fairhurst, G. and Wood, L. (2002). Advice to link designers on link Automatic Repeat reQuest (ARQ). Document, The Internet Society, Reston.

FieldComm Group (2020). HART technology detail [Online]. Available at: https://fieldcommgroup.org/technologies/hart/harttechnology-detail [Accessed 28 July 2020].

Fielding, R.T. and Reschke, J. (2014). Hypertext Transfer Protocol (HTTP/1.1): Message syntax and routing [Online]. Available at: https://rfc-editor. org/rfc/rfc7230.txt.

Frotzscher, A., Wetzker, U., Bauer, M., Rentschler, M., Beyer, M., Elspass, S., Klessig, H. (2014). Requirements and current solutions of wireless communication in industrial automation. In *2014 IEEE International Conference on Communications Workshops*. IEEE, Sydney.

Grossman, E. (2019). Deterministic networking use cases. Document, Internet Engineering Task Force, Fremont.

Hermeto, R.T., Gallais, A., Theoleyre, F. (2017). Scheduling for IEEE 802.15.4-TSCH and slow channel hopping MAC in low power industrial wireless networks: A survey. *Computer Communications*, 114, 84–105 [Online]. Available at: http://www.sciencedirect.com/science/article/pii/S0140366417301147.

Hinden, B. and Deering, D.S.E. (1998). Internet Protocol, version 6 (IPv6) specification [Online]. Available at: https://rfc-editor.org/rfc/rfc2460.txt.

IEEE SA (2016). Standard for Low-Rate Wireless Personal Area Networks (LR-WPANs). Standard 802.15.4-2015, IEEE.

IETF (2020). The Internet of Things [Online]. Available at: https://ietf. org/topics/iot [Accessed 3 August 2020].

Information Science Institute (1981). Internet Protocol [Online]. Available at: https://rfc-editor.org/rfc/rfc791.txt.

Khorov, E., Lyakhov, A., Nasedkin, I., Yusupov, R. (2019). Poster: Fast and reliable alert delivery in Wi-Fi HaLOW sensor networks. In *2019 IFIP Networking Conference (IFIP Networking)*. IFIP, Warsaw.

Koutsiamanis, R.A., Papadopoulos, G.Z., Fafoutis, X., Del Fiore, J.M., Thubert, P., Montavont, N. (2018). From best-effort to deterministic packet delivery for wireless industrial IoT networks. *IEEE Transactions on Industrial Informatics*, 14, 4468–4480.

Koutsiamanis, R.A., Papadopoulos, G.Z., Montavont, N., Thubert, P. (2019). Common ancestor objective functions and parent set DAG metric container extension, draft-ietf-roll-nsa-extension-04. Document, Internet Engineering Task Force, Fremont.

Kumar, S.A.A., Ovsthus, K., Kristensen, L.M. (2014). An industrial perspective on wireless sensor networks – A survey of requirements, protocols, and challenges. *IEEE Communications Surveys Tutorials*, 16(3), 1391–1412.

Lee, G.-W. and Huh, E.-N. (2009). Reliable data transfer using overhearing for implicit ACK. In *2009 ICCAS-SICE*. IEEE, Fukuoka.

Lydon, B. (2011). ISA100 wireless standard review control & instrument manufacturer perspective. White paper, Automation.com, Research Triangle Park, North Carolina.

Mannie, E. (2004). Generalized Multi-Protocol Label Switching (GMPLS) architecture [Online]. Available at: https://rfc-editor.org/rfc/rfc3945.txt.

Martocci, J., Mil, P., Riou, N., Vermeylen, W. (2010). Building automation routing requirements in low-power and lossy networks [Online]. Available at: https://rfc-editor.org/rfc/rfc5867.txt.

Meera, M.S. and Rao, S.N. (2017). A survey of the state of the art of 802.11ah. In *2017 IEEE International Conference on Computational Intelligence and Computing Research (ICCIC)*. IEEE, Tamil Nadu.

Montenegro, G., Kushalnagar, N., Culler, D. (2007). Transmission of IPv6 packets over IEEE 802.15.4 networks. Document, Internet Engineering Task Force, Fremont.

Nixon, M. and Rock, T.R. (2012). A comparison of WirelessHART and ISA100.11a. White paper, Emerson Process Management, 1–36.

O'Dell, M.D., Braden, R.T., Baker, F., Romanow, D.A., Bradner, S.O. (1997). Resource ReSerVation Protocol (RSVP) – Version 1 applicability statement some guidelines on deployment [Online]. Available at: https://rfc-editor.org/rfc/rfc2208.txt.

Palattella, M.R., Accettura, N., Dohler, M., Grieco, L.A., Boggia, G. (2012). Traffic aware scheduling algorithm for reliable low-power multi-hop IEEE 802.15.4e networks. In *2012 IEEE 23rd International Symposium on Personal Indoor and Mobile Radio Communications (PIMRC)*. IEEE, Sydney.

Papadopoulos, G.Z., Koutsiamanis, R., Montavont, N., Thubert, P. (2020). Exploiting packet replication and elimination in complex tracks in LLNs, draft-papadopoulos-raw-pareo-reqs-01. Document, Internet Engineering Task Force, Fremont.

Petersen, S. and Carlsen, S. (2011). Wirelesshart versus isa100.11a: The format war hits the factory floor. *IEEE Industrial Electronics Magazine*, 5(4), 23–34.

Postel, J. (1980). User datagram protocol [Online]. Available at: https://rfc-editor.org/rfc/rfc768.txt.

Qiao, L., Zheng, Z., Cui, W., Wang, L. (2018). A survey on Wi-Fi HaLOW technology for Internet of Things. In *2018 2nd IEEE Conference on Energy Internet and Energy System Integration (EI2)*. IEEE, Beijing.

Rezha, F.P. and Shin, S.Y. (2014). Performance analysis of ISA100.11a under interference from an IEEE 802.11b wireless network. *IEEE Transactions on Industrial Informatics*, 10(2), 919–927.

Shelby, Z., Hartke, K., Bormann, C. (2014). The Constrained Application Protocol (CoAP) [Online]. Available at: https://rfc-editor.org/rfc/rfc7252.txt.

Shi, J., Sha, M., Yang, Z. (2019). Distributed graph routing and scheduling for industrial wireless sensor-actuator networks. *IEEE/ACM Transactions on Networking*, 27(4), 1669–1682.

Thubert, P. and Papadopoulos, G. (2019). Reliable and available wireless problem statement, I-D draft-pthubert-raw-problem-statement-04. Document, Internet Engineering Task Force, Fremont.

Thubert, P., Cavalcanti, D., Vilajosana, X., Schmitt, C. (2019). Reliable and available wireless technologies, I-D draft-thubert-raw-technologies-03. Document, Internet Engineering Task Force, Fremont.

Viswanathan, A., Rosen, E.C., Callon, R. (2001). Multiprotocol label switching architecture [Online]. Available at: https://rfc-editor.org/rfc/rfc3031.txt.

Yazicigil, R.T., Nadeau, P., Richman, D., Juvekar, C., Vaidya, K., Chandrakasan, A.P. (2018). Ultra-fast bit-level frequency-hopping transmitter for securing low-power wireless devices. In *2018 IEEE Radio Frequency Integrated Circuits Symposium*. IEEE, Philadelphia.

Zorbas, D., Kotsiou, V., Théoleyre, F., Papadopoulos, G.Z., Douligeris, C. (2018). Lost: Localized blacklisting aware scheduling algorithm for IEEE 802.15.4-TSCH networks. In *2018 Wireless Days (WD)*, 3–5 April.

6

Adapting Quality of Service of Energy-Harvesting IoT Devices

Matthieu GAUTIER and Olivier BERDER
IRISA, CNRS, University of Rennes 1, Lannion, France

Wireless sensor networks (WSNs) are made up of multiple wireless sensor network nodes that monitor an environment and collect data into one or more locations called sinks. These networks are a key technology in the Internet of Things (IoT) and are essential in many applications such as smart cities, smart factories or precision agriculture. Unlike cellular networks or IEEE 802.11 which rely on a fixed infrastructure, many network topologies are possible for organizing nodes, the simplest being the star network in which all the nodes are located at a distance of one hop from a sink and so send their data into it directly. In the case where the surveillance area is large and where some of the sensor nodes are too distant from the sink for a one-hop communication, a multi-hop network is used.

To deploy large WSNs, one important point is the cost of the nodes, which should be kept low. In addition, so that more applications can be useful WSN should be able to operate over a long period and, generally, the stumbling point of this type of application is energy. Indeed, traditional wireless nodes are supplied by individual batteries, which can only store a limited quantity of energy. The energy consumption of one node, which is directly linked to the Quality of Service (QoS) (detection rate, flow, etc.), is classically defined at deployment at one value, which guarantees the required lifespan. Thus, when a node consumes all the energy initially stored in its battery, it is exhausted, but it is not always possible to replace batteries if the network is dense or deployed in a hostile environment.

Intelligent Security Management and Control in the IoT,
coordinated by Mohamed-Aymen CHALOUF.

A promising solution for increasing the lifespans of WSNs is to allow each node to harvest energy in its environment. In this scenario, each node is equipped with one or more energy harvesters, as well as an energy buffer (battery or capacity) to make it possible to store some of the harvested energy for future use during times when energy is scarce. Various sources of energy are possible, such as light, wind, movement and heat (Vullers *et al.* 2010; Thomas *et al.* 2013). Since sources of energy are generally dynamic and uncontrolled, it is necessary to adapt the energy consumption of the nodes dynamically, by adjusting their QoS to avoid power failures while still maximizing energy efficiency and by guaranteeing that application specifications are met. This task is achieved via Energy Manager (EM), which is responsible for dynamically adapting consumption by nodes, and so their QoS.

This chapter therefore tackles the software and hardware design of energy-harvesting sensor nodes. The typical architecture of sensor nodes considered in these works is illustrated in Figure 6.1. The EM component uses energy information from the node (e.g. the quantity of energy stored and the quantity of energy harvested) to control the node (various tasks, e.g. detection, data processing and communication protocols).

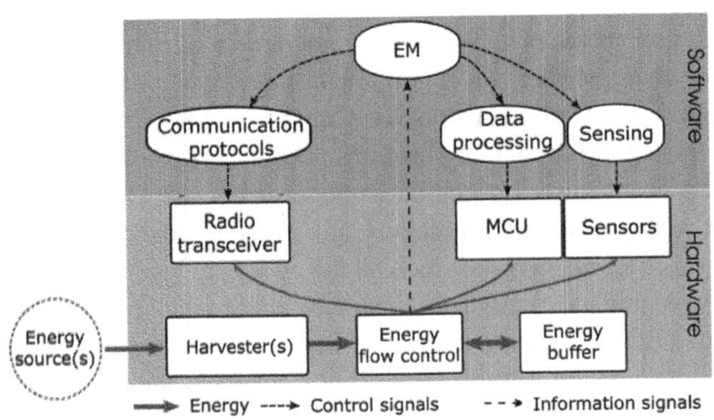

Figure 6.1. *Architecture of an energy-harvesting sensor node. For a color version of this figure, see www.iste.co.uk/chalouf/intelligent.zip*

Section 6.1 presents the problem of managing energy in energy-harvesting sensor networks, as well as a state of the art of energy managers. Sections 6.2 and 6.3 enable us to understand the operation of an energy manager by focusing on two recent and promising examples: the first EM relying on fuzzy logic theory and the second on the principle of Reinforcement Learning (RL). Finally, section 6.4 presents an energy management application using a concrete case: the LoRa long-range radio standard which is part of the Low Power Wide Area Network (LPWAN) family, more and more acclaimed thanks to the compromise that it offers between range and energy consumption.

6.1. Toward the energy autonomy of sensor networks

6.1.1. *Energy harvesting and management*

In energy-harvesting sensor networks, each node is equipped with one or more energy-harvesting devices, allowing the nodes to be supplied entirely by the energy harvested in their environment. To store some of the harvested energy for future use, each node also integrated an energy storing device, which may be a supercapacitor or a battery. This energy buffer has a finite capacity written E_r^{max} and a breakdown threshold written E_r^{fail}, as illustrated in Figure 6.2. The breakdown threshold corresponds to the minimal quantity of energy required for the node to operate, and a breakdown happens if the residual energy passes below this threshold. The E_r^{max} capacity is the maximal quantity of energy that the energy buffer can store. This finite capacity is at the origin of energy wastage. Indeed, if the energy is not immediately used after it is harvested, it is stored in the energy buffer. However, if the energy buffer is full, then the energy harvested will not be stored, and will consequently be wasted.

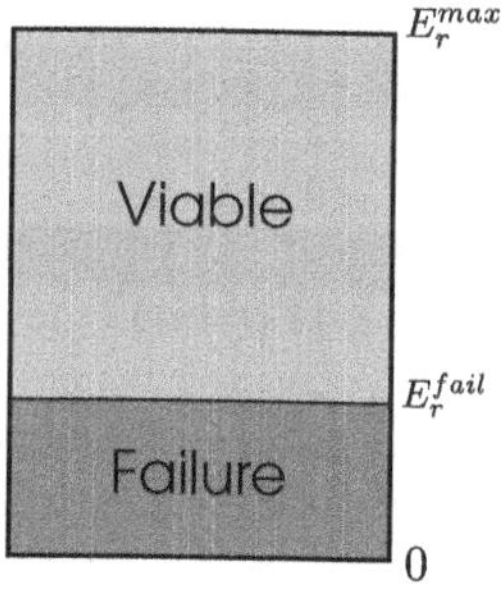

Figure 6.2. *Energy buffer thresholds. For a color version of this figure, see www.iste.co.uk/chalouf/intelligent.zip*

Different sources of energy can be envisaged depending on the environment in which the nodes are deployed and the quantity of energy required per application. The power harvested depends strongly on the type of energy source, as well as the yield of the harvester, that is, on the ratio between the power extracted from the source by the receiver and the power of the source. Table 6.1 gives characteristics of some energy sources (from Vullers *et al.* 2010). As we can see, the power recovered from ambient light outdoors is 10^5 higher than that obtained with energy-harvesting RF (Radio Frequency), demonstrating the great variation in energy available from one source to another. In addition, the sources classically show great temporal variation. This behavior motivates the need for a dynamic adaptation of the nodes' performance. Indeed, the growth in QoS from one application WSN (e.g. the measurement rate and the speed) requires higher energy consumption. One naive solution for obtaining a durable WSN consists of configuring the nodes to a constantly weak QoS, so with an electricity consumption low enough to avoid power failures. This solution presents two main drawbacks: (1) it leads to poor performance since a significant quantity of energy is wasted and (2) a precise knowledge of the energy source is needed to guarantee that no power failure will occur. Another solution consists of carrying out a dynamic performance adaptation, thus allowing the QoS to be maximized while still avoiding power failures.

Source	Power from the source	Harvested power
Ambient light		
Interior	0.1 mW/cm^2	10 μW/cm^2
Exterior	100 mW/cm^2	10 mW/cm^2
Vibration/movement		
Human	0.5 m to 1 Hz 1 m/s^2 to 50 Hz	4 μW/cm^2
Industrial	1 m à 5 Hz 10 m/s^2 to 1 kHz	100 μW/cm^2
Thermal energy		
Humain	20 mW/cm^2	30 μW/cm^2
Industriel	100 mW/cm^2	1-10 mW/cm^2
RF		
GSM base station	0.3 μW/cm^2	0.1 μW/cm^2

Table 6.1. *Characteristics of different sources of energy (from Vullers* et al. *2010)*

In practice, EM is a software program that is mounted on each node and which is run periodically, as illustrated in Figure 6.3. The duration between two executions of the EM is noted T_s. Thus, the residual energy $e_r[k]$ at instant k is defined thus:

$$e_r[k] = e_r[0] + \sum_{i=0}^{k-1} e_h[i] - \sum_{i=0}^{k-1} e_c[i] - \sum_{i=0}^{k-1} e_w[i],\ k = 1, 2, \ldots, \qquad [6.1]$$

where $e_r[0] \in [0, E_r^{max}]$ is the initial residual energy and $e_h[k]$, $e_c[k]$ and $e_w[k]$ are the harvested energy, the energy consumed and the energy wasted, respectively, during the time interval k.

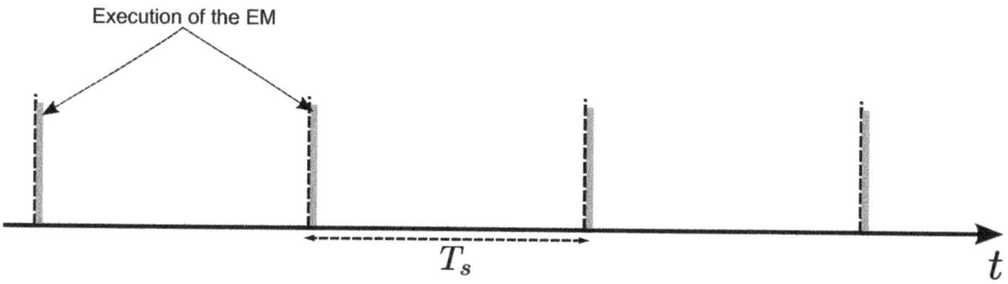

Figure 6.3. *Periodic execution of the EM. For a color version of this figure, see www.iste.co.uk/chalouf/intelligent.zip*

Assuming that the application's QoS can be measured using a utility function $\mathcal{U}(e_c[k],\ k = 0, 1, \dots)$, which depends on the energy consumed (e.g. the packet rate and energy efficiency), the problem of energy management can be formulated by:

$$\begin{aligned} &\underset{e_c[k],\ k=0,1,\dots}{\text{maximize}} \quad \mathcal{U}(e_c[0], e_c[1], \dots) \\ &\text{subject to: } e_r[k] \leqslant E_r^{max},\ k = 1, 2, \dots \\ &\qquad\qquad e_r[k] \geqslant E_r^{fail},\ k = 1, 2, \dots \end{aligned} \qquad [6.2]$$

In this problem, the first constraint is due to the finite capacity of the energy buffer. The second makes it possible to avoid power failures, and satisfying this constraint allows us to expect Energy Neutral Operations (ENO) (Kansal *et al.* 2007). To maximize energy efficiency, the e_c should be chosen so that no energy is wasted, or more formally:

$$e_w[k] = 0,\ k = 1, 2, \dots \qquad [6.3]$$

Satisfying this constraint means meeting the ENO-MAX (Vigorito *et al.* 2007).

However, for some applications, reaching the ENO-MAX is not an appropriate objective. In multi-hop networks for example, where each node generates and relays packets, one node can have a limited packet generation rate, not through the quantity of energy harvested but via its relays which cannot transmit the packets generated. In such a scenario, energy wastage may be needed if the energy buffer is not large enough to store the excess energy harvested. In this study, the emphasis is on star networks, and in such scenarios, energy efficiency, that is, the capacity of an EM to minimize the energy wasted, is pertinent for evaluating the EM. Consequently, the energy efficiency, defined formally thus:

$$\xi = 1.0 - \frac{\sum_k e_w[k]}{\sum_k e_h[k] + e_r[0]}, \qquad [6.4]$$

will be used as a metric for evaluating energy management strategies. We also consider the failure rate D_R, which is defined as the ratio between the time passed in a state of breakdown and the total duration of the application.

6.1.1.1. *Challenges for designing EMs*

Some challenges result from the inability to know (or measure) the energy harvested e_h. As we will see in section 6.1.1.2, some approaches design EMs using predictors that aim to predict the future energy harvested over a finite window of time. However, using predictors requires measurements of the quantity of energy harvested at previous instants $e_h[k-1]$, $e_h[k-2], \ldots$, which are difficult to obtain in practice. Consequently, some approaches rely solely on residual energy to make decisions. Another difficulty in managing energy is precise control of the node's electricity consumption. Indeed, some tasks run by a node have no deterministic energy costs. Communication tasks have for example an energy cost that depends very much on the quality of the channel, a failure in packet transmissions generally leading to retransmissions. Some EMs do not consider this problem; they only calculate an energy budget $e_b[k]$, corresponding to the quantity of energy that the node can dispense during the time interval k; they are not interested in the way in which this energy budget is distributed between the different tasks run by the node. These EMs do not therefore set the QoS directly, but control it indirectly via the energy budget. In this case, another software component is responsible for defining the QoS (e.g. the detection rate) from the energy budget. Other approaches directly define some QoS settings, and so indirectly control the node's energy consumption. In our work, the packet generation rate, written χ_g, is often the parameter for controlling energy consumption. Finally, another challenge is that the wireless sensor nodes are systems that are very constrained in terms of resources. In fact, such devices generally have very little calculating power (a frequency typically less than 16 MHz), memory (typically 2 ko) and storage (typically less than 100 ko), and the proposed algorithms should therefore be adapted to these types of platforms.

6.1.2. *State-of-the-art energy managers*

Many energy management algorithms (Kansal *et al.* 2007; Vigorito *et al.* 2007; Castagnetti *et al.* 2012; Le *et al.* 2012; Yang *et al.* 2013; Hsu *et al.* 2014; Peng and Low 2014; Renner *et al.* 2014; Le *et al.* 2015; Ait Aoudia *et al.* 2017a) have been proposed in recent years to propose efficient adaptation algorithms, adapted to the limited resources provided by the nodes.The different EMs can be classified depending on their need (or not) to predict the quantity of energy that it will be possible to harvest in future, that is, with and without prediction.

6.1.2.1. *Energy managers with prediction*

The general architecture of an EM based on predictions is represented in Figure 6.4(a). It relies on an energy predictor providing the EM with predictions about the energy that can be harvested in the future. For an interval of time k, the energy predictor uses the value of $e_h[k-1]$ to calculate the energy predictions harvested $\hat{e}_h[k]\dots\hat{e}_h[k+K-1]$ on a window of K intervals. The EM then uses these predictions, as well as the current residual energy, to make a decision on the energy budget $e_b[k]$.

The first EM using the prediction approach was introduced in 2007 by Kansal *et al.* (2007). In this schema, an exponentially weighted moving average filter (EWMA) is used to predict the future quantity of energy harvested, and a *duty-cycle* is calculated depending on the gap between the energy predicted and observed. Castagnetti *et al.* (2012) have introduced Closed-Loop Power Manager (CL-PM) that uses two distinct energy management strategies, one for periods where energy is available in the environment, and one for the periods during which the energy harvested is below a fixed threshold. Le *et al.* (2015) have proposed a Wake-up Variation Reduction Power Manager (WVR-PM), one variant of CL-PM that allows a node to store more energy when the energy harvested is substantial in order to obtain a QoS during periods when energy is lacking that is similar to that reached when energy is available. Renner *et al.* (2014) have proposed an algorithm based on the prediction for sensor nodes using a supercapacitor as an energy storage device. The algorithm consists of providing a prediction of the residual energy for a given consumption and a prediction of a given harvesting. The algorithm has been implemented on a network with a dozen nodes with multi-hop topology, and the experiment lasted 4 weeks.

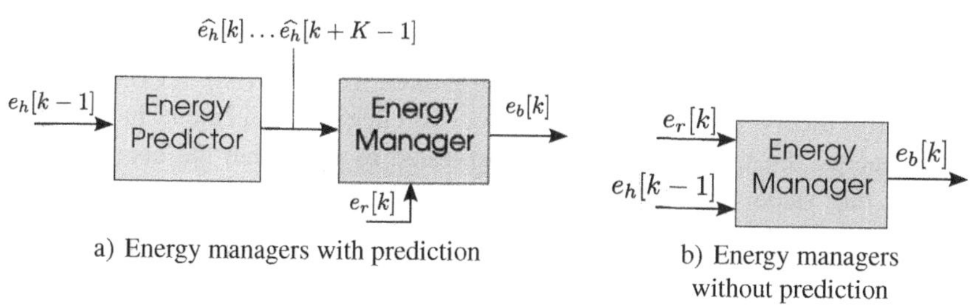

Figure 6.4. *Architecture of energy managers with and without prediction. For a color version of this figure, see www.iste.co.uk/chalouf/intelligent.zip*

The majority of the algorithms proposed assume the possibility of being able to predict the energy harvested (Piorno *et al.* 2009), since the design of precise predictors is an active research domain (Shaikh and Zeadally 2016). However, predictors are generally specific to a type of energy source such as RF (Araújo *et al.* 2014), solar (Bergonzini *et al.* 2009) and wind (Cammarano *et al.* 2012), and some sources are difficult to predict. Thus, techniques that do not use prediction are also proposed.

6.1.2.2. *Energy managers without prediction*

Because the quantity of energy that a sensor can harvest can demonstrate great variation and be difficult to predict, energy predictors can cause significant errors leading to over-use or under-use of the energy harvested (Peng and Low 2014). Thus, some techniques make it possible to calculate an energy budget without using any prediction or any model of the energy source. Figure 6.4(b) gives the structure of this type of EM. LQ-Tracker (Vigorito *et al.* 2007) uses Linear-Quadratic Tracking, a technique taken from adaptive control theory, to adapt the duty-cycle by using only the residual energy as an input. Similarly Le *et al.* (2012) have proposed using a controller Proportional Integral Derivative (PID). With P-FREEN, Peng and Low (2014) have designed an EM that maximizes the duty-cycle of a node where there are inefficiencies in energy storage. The authors have formulated the problem of maximizing the duty-cycle in a nonlinear programming problem and have proposed a set of rules for energy budget attribution using both the energy harvested (actually observed) and the residual energy.

These approaches have demonstrated that it is possible to manage energy efficiently without the additional cost created by prediction mechanisms in terms of memory and calculation quantity. Consequently, our work mainly proposes schemas without prediction. The recent EM P-FREEN is similar to an “all or nothing” policy, that is, the energy budget is either at a minimum value or at a value calculated depending on the energy harvested (to limit energy wastage). This approach can lead to dramatic variations in the allocated budget, ones not generally adapted to WSN applications. In section 6.2, Fuzzyman (Ait Aoudia *et al.* 2016a), an EM based on fuzzy control theory, is presented and allows soft energy management. In addition to residual energy, Fuzzyman uses energy harvested to determine the energy budget. However, this value is difficult to measure in practice. Consequently, section 6.3 presents RLMan (Ait Aoudia *et al.* 2017b, 2018a), an EM based on Reinforcement Learning (RL) in continuous state and action spaces. This EM only requires residual energy as input and is compared to LQ-Tracker (as well as P-FREEN and Fuzzyman), because it is also a schema without prediction using only the residual energy as an input.

6.2. Fuzzyman: use of fuzzy logic

6.2.1. *Design of Fuzzyman*

6.2.1.1. *Fuzzyman architecture*

The aim of Fuzzyman is to calculate an $e_b[k]$ energy budget that the node should provide during the interval of time k depending on the residual energy at the end of the preceding interval $e_r[k-1]$ and the energy harvested during the preceding time interval $e_h[k-1]$. Fuzzyman is therefore run at the end of each interval. Fuzzyman includes three components (Chen and Pham 2000): the blurring unit, the inference engine and the fuzzification unit, as shown in Figure 6.5. Fuzzification consists of transforming physical inputs in a fuzzy set compatible with the inference motor. A fuzzy set is an interval of input data values and a linked membership function. The inference motor defines the rules of actions depending on the fuzzy inputs. Finally, defuzzification makes it possible to link a physical value to the motor's outputs.

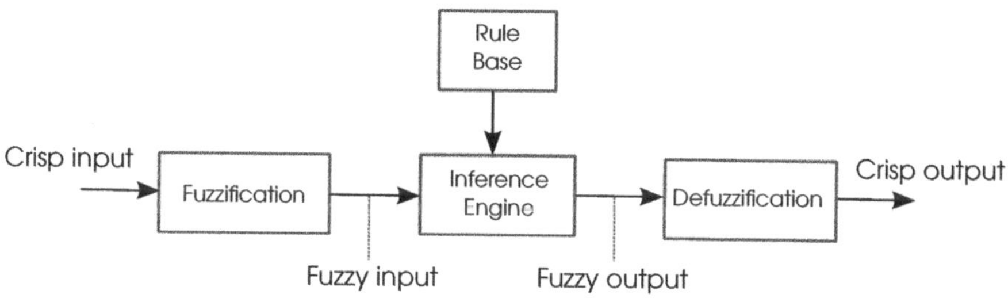

Figure 6.5. *Fuzzyman architecture*

6.2.1.2. *Fuzzification of the controller's inputs*

The first Fuzzyman module is the fuzzification unit that converts each input, that is $e_r[k-1]$ and $e_h[k-1]$, in fuzzy set.

6.2.1.2.1. Fuzzification of recovered energy

To describe the physical value of the harvested energy $e_h \geqslant 0$, two fuzzy sets called WEAK and STRONG are defined. They are associated respectively with the μ_{weak} and μ_{strong} membership functions described in Figure 6.6(b).

If E_b^{min} is the quantity of energy ensuring a minimal QoS for an interval of time, E_h^{weak} considers power leaks P_L and power conversion rates $\eta \in (0,1]$ and is therefore equal to:

$$E_h^{weak} = \frac{E_b^{min}}{\eta} + P_L T_s \qquad [6.5]$$

Thus, if the source is entirely WEAK, then the quantity of energy harvested is insufficient to provide a minimal energy budget E_b^{min}. E_h^{strong} is the threshold at which the harvested energy is considered to be entirely STRONG.

6.2.1.2.2. Fuzzification of residual energy

Three fuzzy sets are defined to describe the residual energy $e_r \in [0, E_r^{max}]$. These sets are called FAIL, EMPTY and FULL and are linked respectively to the membership functions μ_{fail}, μ_{empty} and μ_{full} described in Figure 6.6(b).

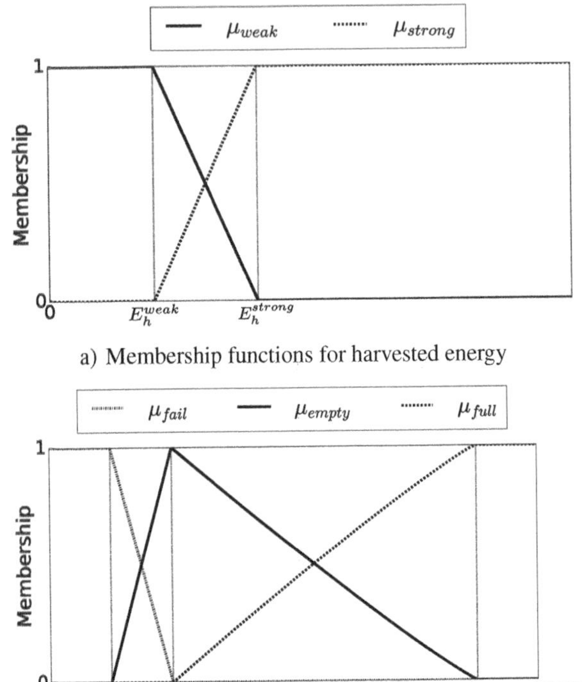

a) Membership functions for harvested energy

b) Membership functions for residual energy

Figure 6.6. *Membership functions used by the fuzzing unit*

E_{res} is defined as the quantity of energy that should be reserved to ensure the minimal QoS when no energy is harvested. E_r^{full} indicates that the node has reserved enough energy E_{res}. E_r^{empty} indicates that all the harvested energy E_{res} has been used. So, $E_{res} = E_r^{full} - E_r^{empty}$. Choosing $E_r^{empty} > E_r^{fail}$ makes it possible to avoid breakdowns when all the energy reserved has been used. Finally, choosing

$E_r^{full} < E_r^{max}$ makes it possible to avoid energy waste when the storage unit is saturated.

6.2.1.3. *Inference motor*

The role of the inference motor is to create control rules depending on the fuzzy inputs for each fuzzification unit. The inference motor's strategy is described by a set of six rules IF-THEN R_i with $i \in \{1 \dots 6\}$ given in Table 6.2. For each interval k, the output of the i rule is written $e_b^i[k]$. When the node has a high risk of electrical failure, it moves into a *state of energy distress* and the quantity of energy needed to ensure minimal functioning for an interval of time in this state is written E_b^{eds} such that $E_b^{eds} < E_b^{min}$. The outputs of rules R_2 and R_3 are given by equation [6.6], and that of the R_6 rule is the energy budget used for the previous interval:

$$e_b^i[k] = E_b^{min} + \mu_{full}(e_r[k-1]) \left(e_h[k-1] - \frac{E_b^{min}}{\eta} - P_L T_s \right) \eta \qquad [6.6]$$

e_h \ e_r	FAIL	EMPTY	FULL
STRONG	E_b^{eds} (R_1)	[6.6] (R_2)	[6.6] (R_3)
WEAK	E_b^{eds} (R_4)	E_b^{min} (R_5)	$e_b[k-1]$ (R_6)

Table 6.2. *Set of rules used by the inference motor*

In equation [6.6], it is assumed that the quantities of energy harvested for two consecutive intervals are similar. This hypothesis is reasonable for a sufficiently short interval duration.

All the rules contain the logical operator AND and apply the following implication :

$$\text{IF } (e_r[k-1] \text{ is } X_r^i) \text{ AND } (e_h[k-1] \text{ is } X_h^i) \text{ THEN } e_b[k] \text{ is } e_b^i[k], \qquad [6.7]$$

where the phrase "x is X" means "x belongs to a fuzzy set X with a membership value $\mu_X(x)$", X_r^i can be FAIL, EMPTY or FULL and X_h^i is STRONG or WEAK. It is important to note that four rules at most can be activated simultaneously by a Fuzzyman execution. The energy management strategy described in Table 6.2 corresponds to the five following scenarios:

1) R_1 and R_4: if the residual energy is FAIL, then the node is in energy distress. In this case, the energy budget is fixed at E_b^{eds};

2) R_5: if the quantity of energy harvested is WEAK and the energy storage device is EMPTY, then the energy budget is fixed at the quantity of energy required to ensure a minimal QoS, that is E_b^{min};

3) R_6: if the quantity of energy harvested is WEAK and the energy storage device is FULL, then the energy budget is unchanged;

4) R_2: if the quantity of energy harvested is STRONG and the energy storage device is still not fully charged, that is $\mu_{full}(e_r[k-1]) < 1$, then some of the energy harvested is used to supply the node, whereas the remainder is stored. The fraction of energy harvested used to supply the node will be at least E_b^{min}, and depends on μ_{full} as indicated by [6.6];

5) R_3: if the quantity of energy harvested is STRONG and the energy storage device is fully charged, that is $\mu_{full}(e_S[k-1]) = 1$, then only the quantity of energy leaked is stored, whereas the remainder is used to supply the node, thus minimizing the risk of energy wastage.

The fuzzy operator AND is implemented by the function min (Chen and Pham 2000). The *activation value* $\alpha_{R_i}[k] \geqslant 0$ for each rule R_i is defined by:

$$\alpha_{R_i}[k] = \mathtt{min}\left\{\mu_{X_r^i}\left(e_r[k-1]\right), \mu_{X_h^i}\left(e_h[k-1]\right)\right\} \qquad [6.8]$$

The R_i rule is *activated* if its activation value is strictly positive. If the rule is not activated, its output value $e_b^i[k]$ is then simply set at 0. At each execution, at least one rule is activated, and so:

$$\sum_{i=1}^{6} \alpha_{R_i}[k] > 0 \qquad [6.9]$$

The activation value of each rule is interpreted as the membership value of the energy budget at the rule's output.

6.2.1.4. *Defuzzing the energy budget*

The defuzzification unit calculates a physical value of the energy balance from the inference motor outputs.

The "center of gravity" formula is the most current (Chen and Pham 2000) to carry out defuzzification.

The physical value of the energy budget is thus calculated by:

$$e_b[k] = \frac{\sum_{i=1}^{6} \alpha_{R_i}[k] e_b^i[k]}{\sum_{i=1}^{6} \alpha_{R_i}[k]} \qquad [6.10]$$

Finally, the complexity of the Fuzzyman algorithm is $O(1)$. It requires few calculations and memory space. Fuzzyman is thus perfectly adapted to sensor network nodes.

6.2.2. *Evaluating Fuzzyman*

6.2.2.1. *Simulation environment*

We use the characteristics of the PowWow platform[1], in which the energy store is a 0.9 F supercapacitor. The nodes are supplied by an interior ambient light. This energy source is diurnal, and its typical light level varies from 1 W/m^2 (Paradiso and Starner 2005) to 10 W/m^2 (Mathúna *et al.* 2008). Simulations have been made using GreenCastalia (Benedetti *et al.* 2013), an *open-source* energy-harvesting simulation environment for simulating Castalia/OMNeT++ (Boulis 2007) networks. The simulated network is formed of a (battery-powered) sink node and four solar-powered nodes. The surface of the solar panel is fixed at 28 cm^2, and its efficiency at 15% (a realistic value compared to current photovoltaic technologies (Green *et al.* 2015)). The simulated PowWow platform uses a TI CC1000 radio that consumes 22.2 mW in receptions, 80.1 mW in transmissions and 0.0006 mW on standby. To evaluate EM performance, only the energy wastage resulting from saturation of the energy storage element is considered, that is, the harvested energy, which cannot be stored because the storage element is full. The power conversion efficiency η is equal to 1 and the leakage power P_L is equal to 0 W. Finally, T_s is fixed at 300 s, and each simulation lasts 31 days (simulated time).

The node's energy consumption is controlled by duty-cycling (Anastasi *et al.* 2009). Each time it wakes up, the node makes a measurement and sends the value obtained thereby to the well. The energy sources provide energy that varies with time and space; the operating cycle of each node is individual and decoupled from the other nodes. This means the synchronous MAC protocols are not adapted to energy-harvesting WSNs, as they require a synchronized operating cycle (Fafoutis and Dragoni 2012). In Ait Aoudia *et al.* (2015), we have shown that a simple MAC CSMA/CA protocol (Carrier Sense Multiple Access with Collision Avoidance) leads to lower energy consumption than the TMAC protocol proposed by Castalia. Consequently, this protocol is used in this work. E_B^{min} is fixed at 0.0221 J which corresponds to a minimal wake up frequency 1/60 Hz.

Fuzzyman is compared to P-FREEN (Peng and Low 2014), an EM without prediction. Three metrics are evaluated: the breakdown rate D_R, corresponding to the time passed out of power, the energy wasted E_W, that is the energy harvested that

1 http://powwow.gforge.inria.fr/.

could not be stored because the energy storage element was full, and the average rate of packets generated $\overline{\chi_g}$.

6.2.2.2. *Energy traces*

GreenCastalia needs energy traces to simulate the energy recovered. In Ait Aoudia *et al.* (2015), a diurnal trace generation algorithm has been proposed. The aim of this algorithm is to evaluate the EMs depending on the characteristics of the energy source that significantly influence their behaviors, and which can be defined by the user as input parameters of the proposed model. In particular, the P_{AVG} parameter makes it possible to define the average power harvested during a day and σ_D makes it possible to control the intensity of fluctuations in the harvested power. For an interior light source with typical power densities ranging from 1 W/m^2 to 10 W/m^2, simulations have been carried out for values of P_{AVG} ranging from 1 W/m^2 to 10 W/m^2, and values of σ_D comprised between 0.1 (low fluctuations from one day to another) and 0.4 (significant fluctuations).

6.2.2.3. *Simulation results*

Fuzzyman and P-FREEN both have an inactivity ratio lower than 0.2% in all simulation scenarios. Consequently, only results in terms of energy wasted and the average rate of packets generated are given in Figure 6.7.

Figure 6.7 has shown the energy wasted E_W when P_{AVG} and σ_D vary. If E_W is similar for both EMs when $P_{AVG} < 2$ W/m^2, Fuzzyman wastes less energy than P-FREEN for higher values of P_{AVG}. The value of σ_D has no impact on the performance of either EM. To evaluate the impact of the energy management efficiency obtained by Fuzzyman, we consider the average rate of packets generated $\overline{P_R}$ which is directly linked to the energy budget allocated by the EMs. Figure 6.7(b) shows the impact of P_{AVG} and σ_D on $\overline{P_R}$. If the average energy budget is similar for both EMs for low values of P_{AVG}, Fuzzyman exceeds P-FREEN for high values of P_{AVG}. In addition, the advantage of Fuzzyman over P-FREEN increases when P_{AVG} increases. Thus, Fuzzyman makes it possible to obtain a maximum gain of 25% compared to P-FREEN (for $P_{AVG} = 10$ mW and $\sigma_D = 0.4$).

6.2.3. **Conclusion**

This section presents the design and evaluation of Fuzzyman, an EM for energy-harvesting nodes based on fuzzy control theory. Fuzzyman is able to provide a high level of efficiency of the harvested energy, while still avoiding breakdowns in power supply. This work presents evaluation of Fuzzyman compared to P-FREEN. Using network simulations supplied by energy traces, we have demonstrated that Fuzzyman exceeds P-FREEN in terms of efficiency of use of the harvested energy.

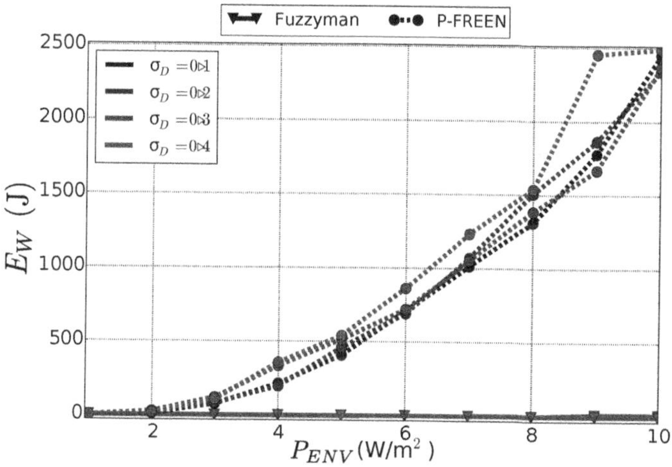

a) Wasted energy E_W

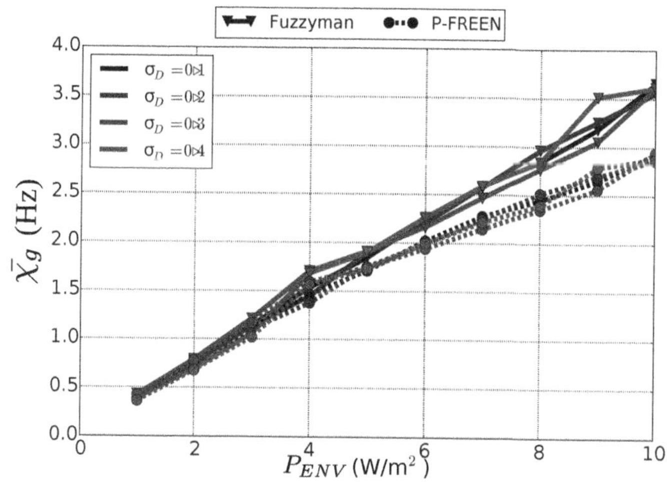

b) Average rate of packets generated $\overline{\chi_g}$

Figure 6.7. *Performance of Fuzzyman and P-FREEN depending on P_{AVG} for different values of σ_D. For a color version of this figure, see www.iste.co.uk/chalouf/intelligent.zip*

However, Fuzzyman determines the energy budget depending on the measurements of the energy harvested. Since this value is difficult to measure precisely in practice (Ait Aoudia *et al.* 2016a; Le Gentil *et al.* 2017), one adaptation of Fuzzyman is proposed in Ait Aoudia *et al.* (2016b). In addition, this consideration has motivated the design of RLMan, presented in section 6.3, an EM based on RL theory.

6.3. RLMan: using reinforcement learning

6.3.1. *Formulating the problem of managing the harvested energy*

In this study, the EM is always run periodically every T_s. The quantity of residual energy is written e_r and the energy storage element is supposed to have a finite capacity written E_r^{max}. The hardware failure threshold, that is, the minimal quantity of residual energy required for the node to operate, is written E_r^{fail}. The role of the node is to send, periodically packets with a $\chi_g \in [X^{min}, X^{max}]$ flow, and the aim of the EM is to adjust dynamically the performance of the node by choosing χ_g. The objective is to maximize the χ_g flow while still keeping the node alive, that is, while avoiding power failures.

In RL, it is supposed that all objectives can be described by maximizing the average cumulative reward, where an R reward is a scalar describing how the node operates. The problem is formulated using an Markov Decision Process (MDP) $\langle \mathcal{S}, \mathcal{A}, \mathcal{T}, \mathcal{R}, \gamma \rangle$ (van Hasselt 2012), with:

– *The state set* $\mathcal{S}$: the state of the node at instant k is defined by the current residual energy $e_r[k]$. Thus, $\mathcal{S} = [E_r^{min}, E_r^{max}]$.

– *The set of actions* $\mathcal{A}$: each time interval, an action, written $A[k]$, is carried out, corresponding to the choice of flow χ_g. Thus, $\mathcal{A} = [X^{min}, X^{max}]$.

– *The transition function* $\mathcal{T}$: the transition function gives the probability of a transition into $e_r[k+1]$ when the action χ_g is carried out in $e_r[k]$ state. Since the state space is continuous, the transition function is a probability density, such that:

$$\Pr\left(e_r[k+1] \in \mathcal{S}[k+1] \mid S[k] = e_r, A[k] = \chi_g\right) = \int_{\mathcal{S}[k+1]} \mathcal{T}\left(S[k] = e_r, A[k] = \chi_g, S\right) dS \quad [6.11]$$

6.3.1.1. *The reward function* $\mathcal{R}$

In this study, the reward is calculated according to χ_g and e_r :

$$R = \phi \chi_g \quad [6.12]$$

with ϕ the normalized residual energy:

$$\phi = \frac{e_r - E_r^{fail}}{E_r^{max} - E_r^{fail}} \quad [6.13]$$

Thus, maximizing the reward means maximizing both the flow and the state of charge of the energy storage element. However, the residual energy depends on the energy consumed and on the energy harvested, and these variables are stochastic. The reward function is defined by:

$$\mathcal{R}(e_r, \chi_g) = \mathbb{E}\left[R[k+1] \mid S[k] = e_r,\ A[k] = \chi_g\right] \quad [6.14]$$

6.3.1.2. *The depreciation factor* γ

The depreciation factor takes its value between 0 and 1. Its function will be detailed below.

The transition function $\mathcal{T}$ describes the dynamic of the environment and is supposed to be unknown. In the same way, by calculating the reward with equation [6.12], its value has a stochastic behavior because it depends on the residual energy. The behavior of a node is defined by a *policy* π, which is a conditional distribution of actions:

$$\pi(\chi_g | e_r) = \Pr\left[A[k] = \chi_g \mid S[k] = e_r\right] \quad [6.15]$$

The goal is to learn a policy that maximizes an objective function. The objective function J is defined as the average of the sum of rewards weighted with respect to a policy π and a distribution of the initial state ρ_0:

$$\begin{aligned} J(\pi) &= \mathbb{E}\left[\sum_{k=1}^{\infty} \gamma^{k-1} R[k] \,\middle|\, \rho_0,\ \pi\right] \\ &= \int_{\mathcal{S}} \rho_\pi(e_r) \int_{\mathcal{A}} \pi(\chi_g | e_r) \mathcal{R}(e_r, \chi_g) d\chi_g de_r \end{aligned} \quad [6.16]$$

with:

$$\rho_\pi(e_r) = \int_{\mathcal{S}} \rho_0(e_r') \sum_{k=1}^{\infty} \gamma^{k-1} \Pr\left[S[k] = e_r \mid S_0 = e_r',\ \pi\right] de_r' \quad [6.17]$$

state distribution e_r under a π policy. In equation [6.16], we can see that choosing a γ value close to 0 leads to a "myopic" evaluation because immediate rewards are favored, whereas choosing a γ value close to 1 leads to a "long-term" evaluation.

Using stochastic policies makes it possible to explore the environment, which is fundamental in RL. Indeed, RL is similar to a "trial and error" learning, and the aim of the algorithm is to discover a good policy arising from its experience with the environment, while still minimizing the quantity of reward "lost" during learning, which leads to a dilemma between exploration (mainly learning from the environment) and exploitation (maximizing the reward using known information).

6.3.2. *RLMan algorithm*

A π policy is evaluated by estimating two value functions associated with this policy. The value function state, written v_π, predicts the future reward if the π policy is used, from a given state. It thus evaluates the "quality" of the states (van Hasselt 2012):

$$v_\pi(e_r) = \mathbb{E}\left[\sum_{i=1}^{\infty} \gamma^{i-1} R_{k+i} \,\middle|\, S[k] = e_r,\ \pi\right] \qquad [6.18]$$

In the same way, the state-action value function associated with carrying out the action $A[k]$ in the state $S[k]$ is written Q_π and evaluates the "quality" of the state-action pairing when the policy π is used (van Hasselt 2012):

$$Q_\pi(e_r, \chi_g) = \mathbb{E}\left[\sum_{i=1}^{\infty} \gamma^{i-1} R_{k+i} \,\middle|\, S[k] = e_r,\ A[k] = \chi_g,\ \pi\right] \qquad [6.19]$$

The proposed EM is an actor-critic algorithm, a family of RL techniques known for its capacity to search for optimal policies with the help of low-variance gradient estimates (Grondman *et al.* 2012). Figure 6.8 shows the relationship between the actor and the critic. The actor updates the π_ψ policy of parameter ψ by a gradient rise of the objective function J [6.16]. A fundamental result for calculating the gradient of J is given by the policy gradient theorem (Sutton *et al.* 2000):

$$\begin{aligned}\nabla_\psi J(\pi_\psi) &= \int_S \rho_{\pi_\psi}(e_r) \int_A Q_{\pi_\psi}(e_r, \chi_g)\, \nabla_\psi\, \pi_\psi(\chi_g \mid e_r) d\chi_g de_r \\ &= \mathbb{E}\left[Q_{\pi_\psi}(e_r, \chi_g)\, \nabla_\psi \log \pi_\psi(\chi_g \mid e_r) \,\middle|\, \rho_{\pi_\psi},\ \pi_\psi\right]\end{aligned} \qquad [6.20]$$

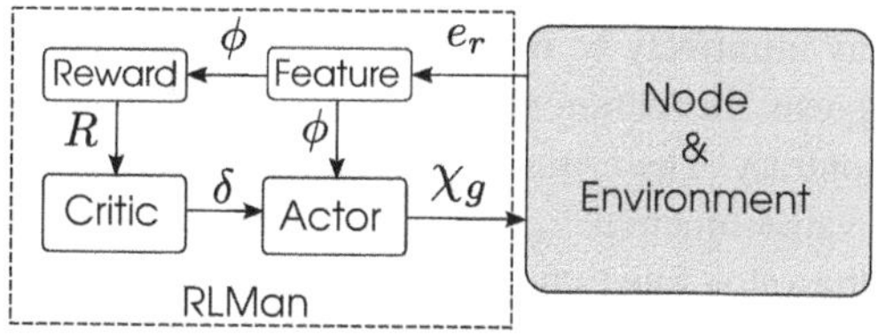

Figure 6.8. *Global RLMan architecture. For a color version of this figure, see www.iste.co.uk/chalouf/intelligent.zip*

This result reduces the calculation to an average and so makes it possible to develop simple algorithms relying on a sample of this average. In this work, a Gaussian policy is used to generate f:

$$\pi_\psi(\chi_g \mid e_r) = \frac{1}{\sigma\sqrt{2\pi}} \exp\left(-\frac{(\chi_g - \mu)^2}{2\sigma^2}\right) \quad [6.21]$$

σ is a constant parameter that controls exploration, and μ is linear with the characteristic ϕ:

$$\mu = \psi\phi \quad [6.22]$$

Defining μ as a linear function of ϕ allows for a reduced memory with only one floating ψ stored. In addition, the excess calculation cost is also reduced because $\nabla_\psi \mu = \phi$ leads to:

$$\nabla_\psi \log \pi_\psi(\chi_g \mid e_r) = \frac{(\chi_g - \mu)}{\sigma^2}\phi \quad [6.23]$$

Other ways of calculating μ can be used, especially by artificial neuron networks for which ψ would be a weight vector of the neurons. These approaches would lead to substantial memory and calculation costs.

By introducing the Temporal Difference (TD-) error defined by:

$$\delta = R[k+1] + \gamma v_{\pi_\psi}(e_r[k+1]) - v_{\pi_\psi}(e_r[k]) \quad [6.24]$$

The policy gradient can be calculated by Grondman *et al.* (2012):

$$\nabla_\psi J(\pi_\psi) = \mathbb{E}\left[\delta \, \nabla_\psi \log \pi_\psi(\chi_g \mid e_r) \,\middle|\, \rho_{\pi_\psi}, \pi_\psi\right] \quad [6.25]$$

The TD-error may intuitively be interpreted as a criticism of the previous action. A positive error suggests this action should be chosen most often, while a negative error suggests the contrary. The criticism calculates the TD error [6.24] and should therefore know the value function v_{π_ψ}. Because the state space is continuous, it is impossible to store the value function for each state. A linear approximation function is therefore also used to estimate the value function:

$$\hat{v}_\theta(e_r) = \theta\phi \qquad [6.26]$$

with ϕ the characteristic [6.13] and θ a parameter of the approximation. The criticism estimates the value function by updating the parameter θ, it is implemented by the TD algorithm (λ) (Sutton 1988) defined by:

$$\nu[k] = \gamma\lambda\nu[k-1] + \phi[k] \qquad [6.27]$$

$$\theta[k] = \theta[k-1] + \alpha\delta[k]\nu[k], \qquad [6.28]$$

where $\alpha \in [0,1]$ is a convergence parameter of the iterations, $\nu[k]$ is a scalar used to attribute a tag to the states visited several iterations earlier and $\lambda \in [0,1)$ is a factor of the weighting rate.

Algorithm 6.1. Energy manager based on reinforcement learning

Require: $e_r[k]$, $R[k]$

1: $\phi[k] = \frac{e_r[k]-E_r^{fail}}{E_r^{max}-E_r^{fail}}$ ▷ Characteristic [6.13]
2: $\delta[k] = R[k] + \gamma\theta[k-1]\phi[k] - \theta[k-1]\phi[k-1]$ ▷ TD error [6.24], [6.26]
3: ▷ Criticism: updating the value function estimate [6.27], [6.28]:
4: $\nu[k] = \gamma\lambda\nu[k-1] + \phi[k]$
5: $\theta[k] = \theta[k-1] + \alpha\delta[k]\nu[k]$
6: ▷ Actor: updating the policy [6.22], [6.23], [6.25]:
7: $\psi[k] = \psi[k-1] + \beta\delta[k]\frac{(f[k-1]-\psi[k-1]\phi[k-1])}{\sigma^2}\phi[k-1]$
8: Setting μ_t in $[X^{min}, X^{max}]$
9: ▷ Generates a new action:
10: $\chi_g[k] \sim \mathcal{N}(\psi[k]\phi[k], \sigma)$
11: Sets $\chi_g[k]$ in $[X^{min}, X^{max}]$
12: **return** $\chi_g[k]$

The Algorithm 6.1 shows the proposed EM. At each execution, the algorithm uses the current residual energy $e_r[k]$ and the reward $R[k]$ calculated with the help of equation [6.12]. The characteristic $\phi[k]$ and the TD error are initially calculated (lines 1 and 2), then the criticism is updated with the help of the TD algorithm(λ) (lines 4 and 5). Then, the actor is updated using the policy gradient theorem at line 7, where $\beta \in [0, 1]$ is a step size parameter. The average of the Gaussian policy is linked to the range of authorized values at line 8. Finally, a frequency is generated by the Gaussian policy of line 10, which will be used in the current time interval. Since the Gaussian distribution is not bounded, it is necessary to set the value generated at the authorized range (line 11).

The algorithm is developed at low memory congestion and reduced complexity, it is therefore adapted to implementation on the WSN nodes.

6.3.3. *Evaluation of RLMan*

RLMan has been evaluated and compared to three state-of-the-art EMs using simulations. The simulated platform is PowWow (Berder and Sentieys 2010). The PowWow platform uses a supercapacitor as an energy storage element, with a maximal tension of 5.2 V and a breakdown tension of 2.8 V. The harvested energy has been simulated by two energy traces taken from real measurements: one of 270 days corresponding to the interior light (Gorlatova *et al.* 2011) and the other of 180 days corresponding to the exterior wind (Jager and Andreas 1996), allowing evaluation of the managers for two different types of energy sources. The role of the node is to measure data, then to send it to a receiver. However, in practice, the quantity of energy consumed by one execution of these tasks varies, for example, because of multiple retransmissions. The quantity of energy required to run one execution has therefore been simulated by a random variable that follows a beta distribution, with the distribution mode equal to the energy consumed if a single transmission attempt is required, written E_c^{typ}, and the maximum set at the energy consumed if five transmission attempts are needed, written E_c^{max}. Table 6.3 gives the values of E_c^{typ} and E_c^{max}, as well as the values of the parameters used when implementing EMs.

RLMan has been compared to P-FREEN, Fuzzyman and LQ-Tracker, three EMs that aim to maximize speed. P-FREEN and Fuzzyman use the harvested energy level in addition to the residual energy level, and have therefore been executed with a perfect knowledge of this value. RLMan and LQ-Tracker use only the residual energy value. The EMs have been evaluated for different capacity values, ranging from 0.5 F to 3.5 F; in fact, capacity has a significant impact on EM behavior, and also on the cost and size of the nodes. The average speed $\overline{\chi_g}$ and the energy efficiency ζ defined by [6.4] have been evaluated.

Common parameters		RLMan		P-FREEN Peng and Low (2014)		Fuzzyman Ait Aoudia *et al.* (2016a)		LQ-Tracker Vigorito *et al.* (2007)	
E_c^{typ}	8.672 mJ	α	0.1	B_{OFL}	$0.95E_r^{max}$	η	1.0	μ	0.001
E_c^{max}	36.0 mJ	β	0.01	η	1.0	E_b^{eds}	$X^{min}T_sE_c^{typ}$	B*	$0.70\ E_r^{max}$
X^{min}	$\frac{1}{300}$ Hz	σ	0.1	–	–	E_b^{min}	$X^{min}T_sE_c^{typ}$	α	0.5
X^{max}	5 Hz	γ	0.9	–	–	E_h^{strong}	$X^{max}T_sE_c^{typ}$	β	1.0
T_s	60 s	λ	0.9	–	–	E_h^{weak}	$X^{min}T_sE_c^{typ}$	–	–

Table 6.3. *Simulation parameter values. Details of parameters of P-FREEN, Fuzzyman and LQ-Tracker are given in the respective publications*

All the EMs succeed in avoiding power failures when they are supplied by an indoor light or outdoor wind. Figure 6.9 gives the simulation results. According to Figures 6.9(c) and 6.9(d), RLMan and LQ-Tracker attain more than 99.9% efficiency, for light and wind and for all the capacity values, even despite the fact that they only use the residual energy as an input. In addition, Figure 6.9(b) shows that, when the node is wind-powered, RLMan surpasses the other EMs in terms of average speed for capacity values. When the node is supplied by an indoor light, RLMan also surpasses the other EMs, except LQ-Tracker for capacity values above 2.8 F. In particular, for a capacity of 0.5 F, the average speed of RLMan is higher by more than 20% compared to LQ-Tracker in the case of interior light, and by almost 70% for outdoor wind. This result is interesting, since using a small capacity makes it possible to reduce the costs and sizes of the nodes.

6.3.4. *Conclusion*

In this section, the problem of maximizing QoS in an energy-harvesting node is formulated with the help of RL theory and an energy management system, RLMan, is presented. RLMan requires only the state of charge of the energy storage element as an input, and uses function approximation to minimize the memory imprint and the calculation time, which makes it practical to implement and makes it adapted to wireless sensor nodes. Simulations have shown the advantages of RLMan in terms of speed and energy efficiency compared to three state-of-the-art EMs, in the case of interior light and exterior wind energy harvesting. The advantage of RLMan is even greater because the energy storage element is small. In addition, this performance is obtained without a need to modify the algorithm in any way. We have identified potential areas of improvement, such as exploring other rewards and other function approximators, which could lead to a faster convergence.

6.4. Toward energy autonomous LoRa nodes

Many technologies aimed at the IoT market have been proposed in recent years, as much by the academic community as by manufacturers. Among these, the standard LoRaWAN, which uses LoRa modulation, enables long-range communications with a limited energy consumption (Le *et al.* 2018). Energy management techniques have mainly been tested on short-range sensor networks with very low energy consumption, and never on long-range networks such as LoRaWAN. In addition, the possibility of exploiting several sources of ambient energy to power the sensor remains little explored. The aim of this section is therefore to design a sensor connected to a LoRaWAN network supplied by several ambient energy-harvesting sources, and which operates entirely autonomously.

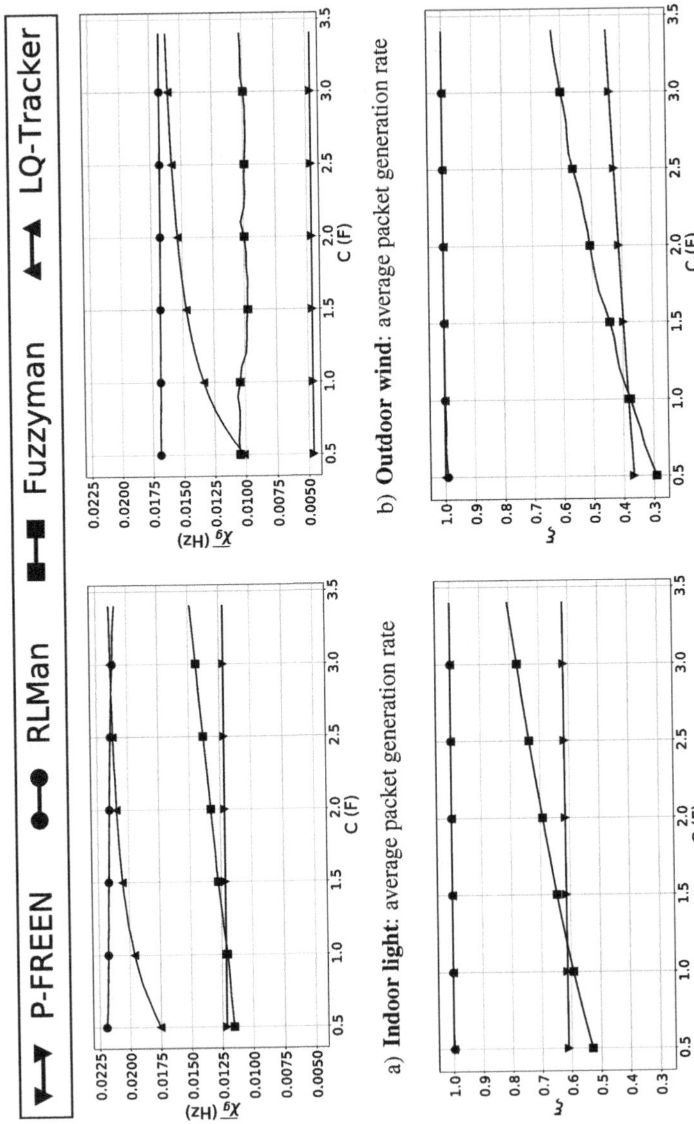

a) **Indoor light**: average packet generation rate

b) **Outdoor wind**: average packet generation rate

c) **Indoor light**: average energy efficiency

d) **Outdoor wind**: average energy efficiency

Figure 6.9. *Average packet generation rate and energy efficiency depending on the value of the capacitor for the indoor light and the outdoor wind*

6.4.1. *Multisource energy-harvesting architecture*

Most existing solutions for energy harvesting to supply LoRa sensors rely on a single energy source (Magno *et al.* 2017; Mabon *et al.* 2019), which potentially reduces the reliability of the sensor. Gleonec *et al.* (2016) and Kazdoghli Lagha *et al.* (2018) present a system making it possible to supply a LoRa sensor by using several energy sources simultaneously. To reduce the cost of implementing the solution, this architecture uses a single power conditioning block Power Management Integrated Component (PIMC) with a circuit making it possible to commute between several heterogenous energy sources. Each source of energy stores the energy harvested in a capacitor. The convertor is successively connected to each condenser, from which it extracts the energy to charge the sensor battery. Despite a limited efficiency, this system makes it possible to reduce the cost of harvesting multisource energy when a large number of energy sources are used.

Figure 6.10 represents the architecture of this circuit. A prototype has been developed to harvest energy from three sources simultaneously and the prototype has been validated with an existing wireless sensor platform, developed by the Wi6labs enterprise, using the LoRa long-range radio protocol. Measurements show that the commutation between energy sources can efficiently combine two sources of energy to increase the autonomy of the device and/or the QoS.

6.4.2. *Applying energy management to LoRa nodes*

Energy management algorithms have mainly been implemented and evaluated using simulation (Gleonec *et al.* 2018a, 2018b). We have evaluated the implementation of these algorithms in real conditions of deploying sensor nodes for IoT. Thus, we have studied the performance of different methods for estimating the energy budget on the LoRa platform.

Figure 6.11 shows the evolution of the performance of four EMs in identical experimental conditions. Detailed in Gleonec *et al.* (2018a), the results show that in this case of use, the choice of algorithm affects the QoS of the system by less than 15%, thus enabling the use of methods with a low complexity of implementation.

6.5. Conclusion

Energy is the main bottleneck for long-term deployment of WSN applications, as the nodes are typically powered by batteries that can store only a finite amount of energy. One promising solution is to allow each node to harvest energy directly from its environment. Since energy sources are dynamic and uncontrollable, it is necessary to adapt continuously the performance of the nodes to optimize the QoS of the applications while still avoiding power failures.

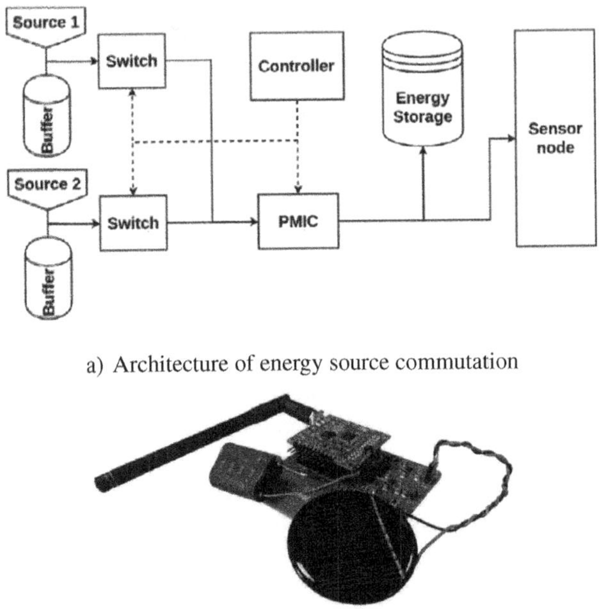

a) Architecture of energy source commutation

b) Prototype linking LoRa and multisource harvesting

Figure 6.10. *Architecture of harvesting multisource energy for LoRa standard. For a color version of this figure, see www.iste.co.uk/chalouf/intelligent.zip*

This chapter makes it possible to understand the operation of an energy manager. Although predictive algorithms, particularly efficient for periodic sources (e.g. external photovoltaic), are now well mastered, new developments are today very promising for more variable environments, which will involve the majority of IoT applications. The first, Fuzzyman, uses fuzzy control theory to calculate an energy budget according to the harvested energy and the residual energy. This algorithm however requires its parameters to be fine-tuned according to the source and the material. In addition, it uses the residual energy, which is a difficult quantity to measure in practice. To free ourselves from these constraints, we have detailed a second algorithm, RLMan, in this chapter, which relies on RL theory. Finally, we have applied energy management to the practical case of the LoRa standard, thus showing the feasibility of supplying power from energy harvesting to develop energy autonomous applications.

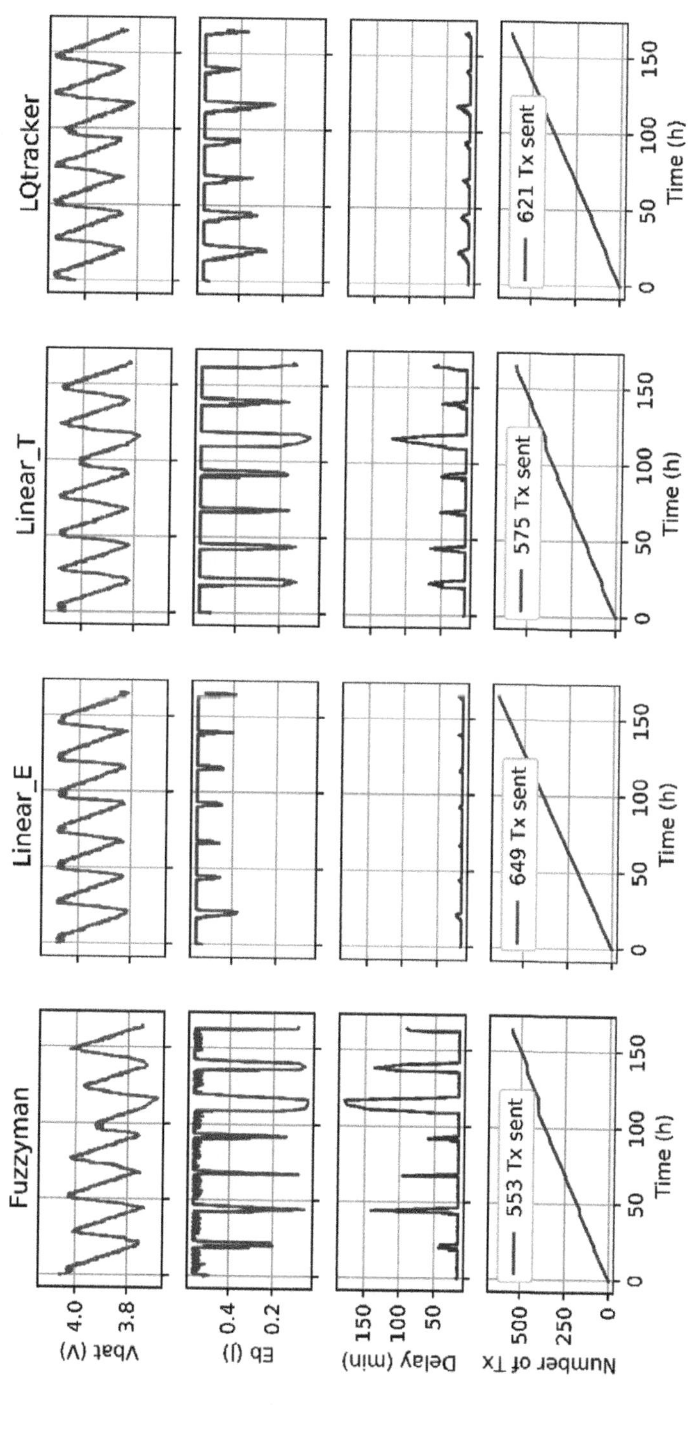

Figure 6.11. *Performance of different EMs over 7 days of execution. For a color version of this figure, see www.iste.co.uk/chalouf/intelligent.zip*

6.6. References

Ait Aoudia, F., Gautier, M., Berder, O. (2015). GRAPMAN gradual power manager for consistent throughput of energy harvesting wireless sensor nodes. In *IEEE International Symposium on Personal, Indoor, and Mobile Radio Communications (PIMRC 15)*. IEEE, Hong Kong.

Ait Aoudia, F., Gautier, M., Berder, O. (2016a). Fuzzy power management for energy harvesting wireless sensor nodes. In *IEEE International Conference on Communications (ICC 16)*. IEEE, Kuala Lumpur.

Ait Aoudia, F., Gautier, M., Le Gentil, M., Berder, O. (2016b). Demo abstract: How fuzzy logic can enhance energy management in autonomous wireless sensor nodes? In *Conference on Design and Architectures for Signal and Image Processing (DASIP), Demo Night*. IEEE, Porto.

Ait Aoudia, F., Gautier, M., Berder, O. (2017a). Distributed computation of fair packet rates in energy harvesting wireless sensor networks. *IEEE Wireless Communications Letters*, 6(5), 626–629.

Ait Aoudia, F., Gautier, M., Berder, O. (2017b). Learning to survive: Achieving energy neutrality in wireless sensor networks using reinforcement learning. In *IEEE International Conference on Communications (ICC)*. IEEE, Paris.

Ait Aoudia, F., Gautier, M., Berder, O. (2018a). RLMan: An energy manager based on reinforcement learning for energy harvesting wireless sensor networks. *IEEE Transactions on Green Communications and Networking*, 2(2), 408–417.

Ait Aoudia, F., Gautier, M., Magno, M., Berder, O., Benini, L. (2018b). Leveraging energy harvesting and wake-up receivers for long-term wireless sensor networks. *Sensors*, 5, 1–27.

Anastasi, G., Conti, M., Francesco, M.D., Passarella, A. (2009). Energy conservation in wireless sensor networks. A survey. *Ad Hoc Networks*, 7(3), 537–568.

Araújo, G.M.D., Pinto, A.R., Kaiser, J., Becker, L.B. (2014). *Genetic Machine Learning Approach for Link Quality Prediction in Mobile Wireless Sensor Networks*. Springer, Berlin, Heidelberg.

Benedetti, D., Petrioli, C., Spenza, D. (2013). GreenCastalia: An energy-harvesting-enabled framework for the Castalia simulator. In *ACM International Workshop on Energy Neutral Sensing Systems*. ENSSYS, Rome.

Berder, O. and Sentieys, O. (2010). PowWow: Power optimized hardware/software framework for wireless motes. In *International Conference on Architecture of Computing Systems*. ARCS, Hannover.

Bergonzini, C., Brunelli, D., Benini, L. (2009). Algorithms for harvested energy prediction in batteryless wireless sensor networks. In *International Workshop on Advances Sensors and Interfaces*. IEEE, Trani.

Boulis, A. (2007). Castalia: Revealing pitfalls in designing distributed algorithms in WSN. In *ACM International Conference on Embedded Networked Sensor Systems*. SENSYS, Sydney.

Cammarano, A., Petrioli, C., Spenza, D. (2012). Pro-Energy: A novel energy prediction model for solar and wind energy-harvesting wireless sensor networks. In *IEEE International Conference on Mobile Adhoc and Sensor Systems (MASS)*. IEEE, Las Vegas.

Castagnetti, A., Pegatoquet, A., Belleudy, C., Auguin, M. (2012). A framework for modeling and simulating energy harvesting WSN nodes with efficient power management policies. *EURASIP Journal on Embedded Systems*, 8, 1–20.

Chen, G. and Pham, T.T. (2000). *Introduction to Fuzzy Sets, Fuzzy Logic, and Fuzzy Control Systems*. CRC Press, Boca Raton.

Fafoutis, X. and Dragoni, N. (2012). Analytical comparison of MAC schemes for energy harvesting – Wireless sensor networks. In *International Conference on Networked Sensing Systems*. INSS, Anvers.

Gleonec, P.-D., Ardouin, J., Gautier, M., Berder, O. (2016). Architecture of multi-source energy harvesting for IoT nodes. In *IEEE Online Conference on Green Communications (Online GreenComm 2016)*.

Gleonec, P.-D., Ardouin, J., Gautier, M., Berder, O. (2018a). A real-world evaluation of energy budget estimation algorithms for autonomous long range IoT nodes. In *International Conference on Telecommunications*. IEEE, Saint-Malo.

Gleonec, P.-D., Ardouin, J., Gautier, M., Berder, O. (2018b). Demo abstract: Multi-source energy management for long range IoT nodes. In *International Conference on Telecommunications*. IEEE, Saint-Malo.

Gorlatova, M., Wallwater, A., Zussman, G. (2011). Networking low-power energy harvesting devices: Measurements and algorithms. In *IEEE International Conference on Computer Communications (INFOCOM)*. IEEE, Shanghai.

Green, M.A., Emery, K., Hishikawa, Y., Warta, W., Dunlop, E.D. (2015). Solar cell efficiency tables (version 45). *Progress in Photovoltaics: Research and Applications*, 23(1), 1–9.

Grondman, I., Busoniu, L., Lopes, G.A.D., Babuska, R. (2012). A survey of actor-critic reinforcement learning: Standard and natural policy gradients. *IEEE Transactions on Systems, Man, and Cybernetics, Part C (Applications and Reviews)*, 42(6), 1291–1307.

van Hasselt, H. (2012). Reinforcement learning in continuous state and action spaces. In *Reinforcement Learning*, Wiering, M., van Otterlo, M. (eds). Springer, Berlin, Heidelberg.

Hsu, R.C., Liu, C.T., Wang, H.L. (2014). A reinforcement learning-based ToD provisioning dynamic power management for sustainable operation of energy harvesting wireless sensor node. *IEEE Transactions on Emerging Topics in Computing*, 2(2), 181–191.

Jager, D. and Andreas, A. (1996). NREL National Wind Technology Center (NWTC) [Online]. Available at: http://www.nrel.gov/midc/nwtc_m2/ [Accessed 11 January 2022].

Kansal, A., Hsu, J., Zahedi, S., Srivastava, M.B. (2007). Power management in energy harvesting sensor networks. *ACM Transactions on Embedded Computing Systems*, 6(4), 32-es.

Kazdoghli Lagha, M., Ait Aoudia, F., Gautier, M., Berder, O. (2018). Feature selection framework for multi-source energy harvesting wireless sensor networks. In *IEEE Vehicular Technology Conference (VTC-Spring 18)*. IEEE, Porto.

Le, T.N., Sentieys, O., Berder, O., Pegatoquet, A., Belleudy, C. (2012). Power manager with PID controller in energy harvesting wireless sensor networks. In *IEEE International Conference on Green Computing and Communications (GreenCom)*. IEEE, Besançon.

Le, T.N., Pegatoquet, A., Berder, O., Sentieys, O. (2015). Energy-efficient power manager and MAC protocol for multi-hop wireless sensor networks powered by periodic energy harvesting sources. *IEEE Sensors Journal*, 15(12), 7208–7220.

Le, X.-C., Vrigneau, B., Gautier, M., Mabon, M., Berder, O. (2018). Energy/reliability trade-off of LoRa communications over fading channels. In *International Conference on Telecommunications*. IEEE, Saint-Malo.

Le Gentil, M., Ait Aoudia, F., Gautier, M., Berder, O. (2017). Demo abstract: Online QoS adaptation using fuzzy control for indoor light energy harvesting in wireless sensor networks. In *IEEE International Workshop on Signal Processing Systems (SiPS)*. IEEE, Lorient.

Mabon, M., Gautier, M., Vrigneau, B., Le Gentil, M., Berder, O. (2019). The smaller the better: Designing solar energy harvesting sensor nodes for long-range monitoring. *Wireless Communications and Mobile Computing*, 2019, 1–11.

Magno, M., Ait Aoudia, F., Gautier, M., Berder, O., Benini, L. (2017). WULoRa: An energy efficient IoT end-node for energy harvesting and heterogeneous communication. In *IEEE/ACM Design, Automation & Test in Europe Conference & Exhibition (DATE)*. IEEE/ACM, Lausanne.

Mathúna, C.Ó., O'Donnell, T., Martinez-Catala, R.V., Rohan, J., O'Flynn, B. (2008). Energy scavenging for long-term deployable wireless sensor networks. *Talanta*, 75(3), 613–623.

Paradiso, J. and Starner, T. (2005). Energy scavenging for mobile and wireless electronics. *IEEE Pervasive Computing*, 4(1), 18–27.

Peng, S. and Low, C.P. (2014). Prediction free energy neutral power management for energy harvesting wireless sensor nodes. *Ad Hoc Networks*, 13, 351–367.

Piorno, J.R., Bergonzini, C., Atienza, D., Rosing, T.S. (2009). Prediction and management in energy harvested wireless sensor nodes. In *International Conference on Wireless Communication, Vehicular Technology, Information Theory and Aerospace Electronic Systems Technology (Wireless VITAE)*. IEEE, Aalborg.

Renner, C., Unterschütz, S., Turau, V., Römer, K. (2014). Perpetual data collection with energy-harvesting sensor networks. *ACM Transactions on Sensor Networks*, 11(1), 1–45.

Shaikh, F.K. and Zeadally, S. (2016). Energy harvesting in wireless sensor networks: A comprehensive review. *Renewable and Sustainable Energy Reviews*, 55, 1041–1054.

Sutton, R.S. (1988). Learning to predict by the methods of temporal differences. *Machine Learning*, 3(1), 9–44.

Sutton, R.S., McAllester, D.A., Singh, S.P., Mansour, Y. (2000). Policy gradient methods for reinforcement learning with function approximation. *Advances in Neural Information Processing Systems 12*, 99, 1057–1063.

Thomas, Y.R., Picot, M., Carer, A., Berder, O., Sentieys, O., Barrière, F. (2013). A single sediment-microbial fuel cell powering a wireless telecommunication system. *Journal of Power Sources*, 241, 703–708.

Vigorito, C.M., Ganesan, D., Barto, A.G. (2007). Adaptive control of duty cycling in energy-harvesting wireless sensor networks. In *IEEE Communications Society Conference on Sensor and Ad Hoc Communications and Networks (SECON)*. IEEE, San Diego.

Vullers, R.J.M., van Schaijk, R., Visser, H.J., Penders, J., Hoof, C.V. (2010). Energy harvesting for autonomous wireless sensor networks. *IEEE Solid-State Circuits Magazine*, 2(2), 29–38.

Yang, S., Yang, X., McCann, J.A., Zhang, T., Liu, G., Liu, Z. (2013). Distributed networking in autonomic solar powered wireless sensor networks. *IEEE Journal on Selected Areas in Communications*, 31(12), 750–761.

7

Adapting Access Control for IoT Security

Ahmad KHALIL, Nader MBAREK and Olivier TOGNI
LIB, Bourgogne Franche-Comté University, Dijon, France

7.1. Introduction

The Internet of Things (IoT) is a new paradigm making it possible to improve human quality of life while still ensuring that advanced intelligent services are used in our daily lives by means of interconnecting different objects (telephones, sensors, etc.) and entities (human, animal, etc.). These services create new security challenges in this type of environment, which require specific measures to improve the security level and in particular to protect information that transits via IoT infrastructures. One of the major challenges of the IoT involves the limitations in terms of energy capacities, processing capacities and the capacities of its components' memory. Consequently, adaptation of different security mechanisms is a requirement in this IoT environment to meet these limitations. In this chapter, we define in section 7.2 the different security services in the IoT. Then, section 7.3 presents the most widespread access control techniques. In section 7.4, we study research on access control in the IoT. This work relates to the adaptation of access control models and the adaptation of access control systems and technologies by considering the characteristics of the IoT. Then, we propose in section 7.5 our adaptation of access control in the IoT. This adaptation is based on an access control architecture and mechanism called IoT-MAAC (IoT-Multiple Attribute Access Control) taking account of the different attributes specific to the IoT environment, including the trust level of IoT objects, calculated via a fuzzy logic system. Finally,

Intelligent Security Management and Control in the IoT,
coordinated by Mohamed-Aymen CHALOUF.

section 7.6 presents conclusions and perspectives on the adaptation of access control in the IoT environment.

7.2. Defining security services in the IoT

In this section, we present the different security services to consider in all IoT infrastructures in order to respect the security and privacy of IoT service users. Particular attention will be paid, in section 7.3, to the access control service, which is the subject of this chapter.

7.2.1. *Identification and authentication in the IoT*

Identification consists of establishing the identity of the IoT service user. In principle, it relies on using an identifier attributed individually to a user. Authentication follows identification and allows the user to prove their identity. The user uses an authenticator or a secret code known only to them. Nevertheless, authentication does not mean right of access, it is access control that grants this privilege if authentication has succeeded (ITU-T 1991). Identification and authentication mechanisms can provide advantages to the IoT environment. Thus, this environment, through these identification and authentication mechanisms, integrates robust devices capable of reducing the risks of intrusion and avoiding user privacy violations (Li 2017).

In addition, traditional identification and authentication mechanisms should be adapted to respect the demands of the IoT environment in terms of scalability, the large number of entities, etc. Several organizations use digital certificates based on PKI (Public Key Infrastructure) for operations identifying and authenticating peripheral devices (Joshi 2017). However, adaptations should be made to consider this type of solution in the IoT. First, PKI infrastructures should be able to support effectively the process of transmitting digital certificates in large quantities and at great speed. Then, digital certificates have a limited lifespan, which means that they have an expiry date. In the context of the IoT environment, some usage cases can require short-term certificates, while many others require longer term certificates. Thus, a longer term certificate is required when a peripheral IoT device requires a long-term authentication. Consequently, those in charge of IoT projects should carefully determine the lifespan required for digital certificates and determine the advantages and disadvantages of this. On the other hand, certificates, since they are considered critical elements with a lifecycle, should be managed effectively. Indeed, manually managing certificate monitoring is not feasible in an IoT environment. Thus, PHI infrastructure should be linked to certificate management providers with a scalable platform. These platforms should then be able to manage cases of use in the IoT (Joshi 2017). Other solutions and technologies can also be used for authentication in the IoT such as SAML (Security Assertion Markup Language). This is a standard, developed by OASIS (Organization for the Advancement of Structured Information Standards), defining a protocol for exchanging information linked to security. SAML, based on XML language, proposes a unique authentication on the Web. In addition, the SAML standard is modular, thanks to authentication messages that can be routed using a number of communication protocols (HTTP: Hypertext Transfer Protocol, SOAP: Simple Object Access Protocol, etc.) (OASIS 2019a). Finally, this standard is compatible with other access control standards such as XACML (eXten-sible Access Control Markup Language) (OASIS 2019b).

7.2.2. *Access control in the IoT*

Access control makes it possible to combat unauthorized use of an IoT resource. To achieve this control, a list of entities authorized to access a resource, with their levels of access, is defined, in conformity with a security policy. Thus, a service is offered to put in place different types of access to resources (reading, writing, modification and deletion of information and executing a task). Access control is based on one or more elements (an access control list, a hierarchical or distribution structure matrix, etc.) via databases kept by authorization centers or by the resource. These databases include authentication information (passwords, security labels, etc.) (ITU-T 1991). In the context of access control in the IoT, two types of entity are important: data holders (IoT service users) and data collectors (IoT objects) that send data or receive commands. These two types of entity should be authenticated mutually (Balte *et al.* 2015).

Several models exist in the literature to ensure the access control service. Access control solutions based on the RBAC (Role-Based Access Control) model use three elements to make access control decisions: the user, the role and the permissions. Authorization is carried out depending on the role of the user. On the other hand, ABAC (Attribute-Based Access Control) is another model that takes the access control decision based on several different attributes of the entity requesting access to a resource. The attributes may belong to the entity (identity) or can depend on the environment (request time) (Maw *et al.* 2014). Other access control models exist such as capBAC (capability Based Access Control) offering tickets or tokens to authorized entities, orBAC (organizational Based Access Control) using context data (historical, spatial, temporal, declared by the user, etc.) and MAC (Mandatory Access Control) that allocates labels to objects to specify access control rules (Maw *et al.* 2014).

7.2.3. *Confidentiality in the IoT*

The confidentiality service provides protection against traffic analysis by unauthorized entities and against divulging data flows. To assure this security service, data encryption is the most adapted security mechanism. Encryption can be carried out using a symmetrical system (a secret key) or an asymmetrical system (a public key). The symmetrical encryption system involves knowing the secret key, which allows encryption and decryption. For the asymmetrical encryption system, knowing the public encryption key, for all the entities, does not involves knowing the corresponding private key for decryption. In addition to encryption mechanisms, there needs to be a key management mechanism to exchange these keys between communicating entities (ITU-T 1991).

In the IoT environment, several points should be considered when implementing a confidentiality service, especially during the process of exchanging the keys used for encryption. For this, extendibility is an important characteristic given the large number of connected objects. Indeed, the number of entities that can be involved in the process of exchanging keys can be limited using traditional systems, whereas new entities can be involved after the first exchange of keys and new objects can be integrated into the IoT environment after services have been initiated. On the other hand, scalability is an important characteristic to take into consideration. Indeed, after the involvement of new entities in the process of exchanging keys in the Internet of objects, the volume of cryptographic data to be stored on the objects is becoming ever greater, while IoT objects have restrictions in terms of storage and processing capacity (Abdemeziem 2016).

7.2.4. *Integrity in the IoT*

Integrity is a security service that covers two large aspects of the context of the IoT: data integrity and object integrity. Data integrity aims to ensure that the data exchanged in an IoT environment have not been modified or destroyed in an unauthorized manner during their routing. It is needed to provide a reliable service and to ensure that the information collected and the commands received by the objects are legitimate. Data integrity verification involves two processes, one at the transmitter and one at the receiver. The transmitting entity adds control information (control code per block such as BCC (Block Check Character) or cryptographic control value such as hashing) depending on the data sent. The receiving entity generates the same information based on the data and compares it to the data received to determine if these data have been modified during routing in the IoT environment (ITU-T 1991).

The second type of integrity in the IoT context is the type that involves objects. Thus, object integrity is needed since objects in the IoT can be deployed in non-trusted environments and can be attacked physically to modify, for example, the code when it is run on the objects. This second integrity service for assurance in the IoT makes it possible to detect and prevent the modifications made to the operating system and the configuration of objects. Object integrity in the IoT also makes it possible to lock and eliminate peripheral devices that do not conform. To implement this type of integrity, a digital fingerprint of the object in question is used to compare the existing data on the object and the data that should exist on it.

7.2.5. *Non-repudiation in the IoT*

A non-repudiation security service makes it possible not to deny participation in exchanges by a communicating entity. This service can appear in many forms. The first form is non-repudiation with proof of origin, where the recipient receives proof of the data's origin. This proof may be formed of a digital signature using asymmetrical encryption applied to the hash value of the exchanged data. The second form is non-repudiation with proof of data submission, where the transmitter receives proof in the form of acknowledgment of receipt for example (ITU-T 1991). This non-repudiation security service is necessary in the IoT to provide proof that data have been sent by the things, and also a proof of the order of execution has been sent by IoT service users. This makes it possible to follow and save all events that have taken place in an IoT environment in trace files.

7.2.6. *Availability in the IoT*

Availability involves the opportunity to access and use resources on demand, for an entity that has been authorized following authentication and access control. Availability is a security service since a service that is unavailable following a DoS (Denial of Service) type attack for example is a non-secure service and may be compromised at any moment (Mosenia and Jha 2017). Availability in the IoT is essential to provide an environment that is connected to the Internet and fully operational. In the context of the IoT, this service integrates on the one hand the availability of devices (objects, gateways) making it possible to collect data without interruption. On the other hand, it thus ensures the availability of the IoT service offered to users. This second type of availability is determined by the configuration of the IoT environment, hence the need to make appropriate choices of management and administration protocols and to protect against DoS (or DDOS: Distributed DoS) type attacks. The services offered in the IoT should be available continuously while still considering the criticality of some of these services (Mosenia and Jha

2017). In this context, permanent 24-hour availability 365 days per year is needed for critical services in the IoT, such as some applications in the e-health domain.

7.3. Access control technologies

Different technologies can be used to ensure access control through authentication or identification in the IoT environment. Among these technologies, two standards, XACML (eXtensible Access Control Markup Language) and SAML, jointly make it possible to control access within this type of environment.

XACML, proposed by the OASIS standardization body, is an access control policy language providing two-directional communication between the access request and the corresponding response. This policy language is very widespread and can operate with different access control models (such as RBAC, ABAC, etc.) (Atlam *et al.* 2018). This language is compatible with several standards, among which we can mention SAML and LDAP (Lightweight Directory Access Protocol) (Atlam *et al.* 2018).

The different components and modules of the XACML standard are the following (see Figure 7.1):

– PAP (Policy Administration Point): the policy administration point is an interface for defining policies by making them available to the policy decision point module;

– PEP (Policy Enforcement Point): the policy enforcement point receives the access request from the service user and applies the access decision received by the intermediary of the context handler (Context Handler);

– PDP (Policy Decision Point): the policy decision point evaluates the access request in relation to the applicable policies and resends the access decision to the PEP via the context handler (Context Handler);

– Context Handler: the context handler translates the access requests to transmit them to the PDP. It acts as a bridge between the PEP and the PDP and it is responsible for recovering the attribute values needed to evaluate the corresponding access request policy;

– PIP (Policy Information Point): the information point on the policy provides the attribute values making it possible to take the decision. It interacts directly with the context handler and the subjects of the access control.

An XACML strategy includes policies and each policy is formed by a set of rules regrouped with the help of a rule combination algorithm. These rules are formed of three elements: a target, an effect and a condition. On the one hand, the target makes it possible to decide the rule to consider. On the other hand, the effect is the result to be provided (accept or refuse) when a rule is satisfied, while the condition is the function to run when a rule is applicable (Atlam *et al.* 2018; XACML 2019). In this context, the XACML engine evaluates a given XACML request in relation to multiple rules independently. The XACML engine is a software component that requires two inputs to find a result as an output. The two entries are XACML policies and the access request, while the result concerns the access decision. The result for a data access request may be presented in one of the following forms: authorization, refusal and not applicable (XACML 2019).

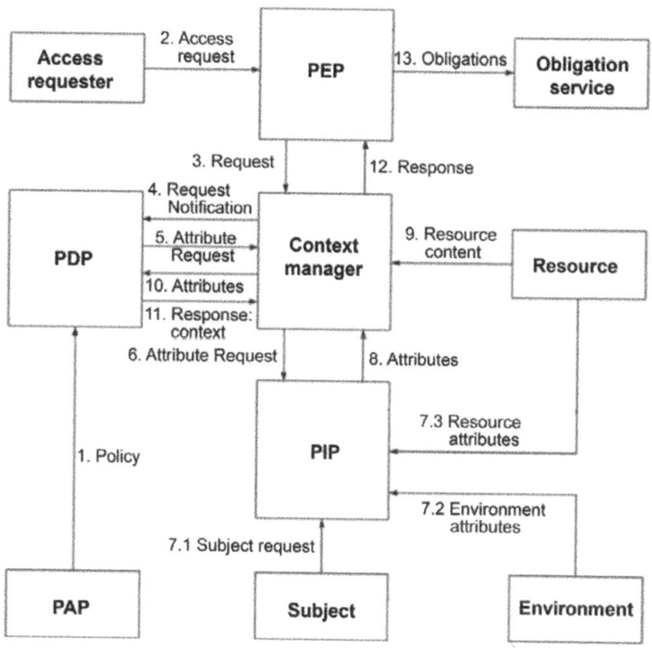

Figure 7.1. *XACML components and modules (XACML 2019)*

To assure access control, an authentication stage is needed. This security service can be provided using SAML. SAML (SAML 2019) language, standardized by OASIS, is broadly used to implement a unique inter domain connection (SSO: Single Sign On). It can be mapped onto other, different protocols such as SOAP, HTTP, etc. Indeed, SAML is a standard permitting the exchange of authentication and authorization data between partners, in particular between an identity provider (IdP) and a service provider (SP). The IdP creates and manages subjects' identity data and provides authentication to different SPs. The SP is an entity that provides services for authenticated subjects. The IdP can request information from the subjects (user names, passwords, etc.) to authenticate them. To authenticate a subject, SAML uses the notion of assertion that is transmitted from the IdP to the SPs. The SAML assertions make it possible to diffuse information on security that can be used for diverse ends. One of the most important objectives is contributing to access control decisions. SAML does not specify how this information should be used or how the access control strategies should be handled. Indeed, SAML does not specify the authentication method at the IdP. The IdP can use a user name and a password, or another form of authentication, including multi-factor authentication.

SAML and XACML are used jointly to assure authentication followed by access control (see Figure 7.2). In this context, the SAML standard defines the method of mapping different attributes used by SAML when authenticating in XACML attributes to be used during access control via the XACML attribute profile. This attribute profile also defines a new type of decision authorization request specially designed for use in an XACML environment.

The XACML attribute profile extends use of the SAML protocol to the XACML protocol by allowing the SAML protocol to be used to extract an XACML protocol that can be applied to take an access decision, and also to allow the SAML assertion to be used to wrap an XACML policy or even to exchange the information needed for identification and authentication.

Combined use of SAML and XACML generally involves the following stages:

– the PEP receives a request for access to a resource and obtains SAML assertions containing information on the access requesters and resources involved in the access request. These assertions can accompany the access request or they can be obtained directly from an SAML authority;

– the PEP obtains other information pertinent to the access request, such as the time, the date, the location and the resource properties;

– the PEP presents all the information obtained thanks to, among others, SAML and the PDP to decide whether access should be authorized;

– the PDP obtains all the policies pertinent to the access request and evaluates them;

– the PDP informs the PEP of the decision on access control;

– the PEP applies the decision by authorizing the requested access or by indicating that access is not authorized.

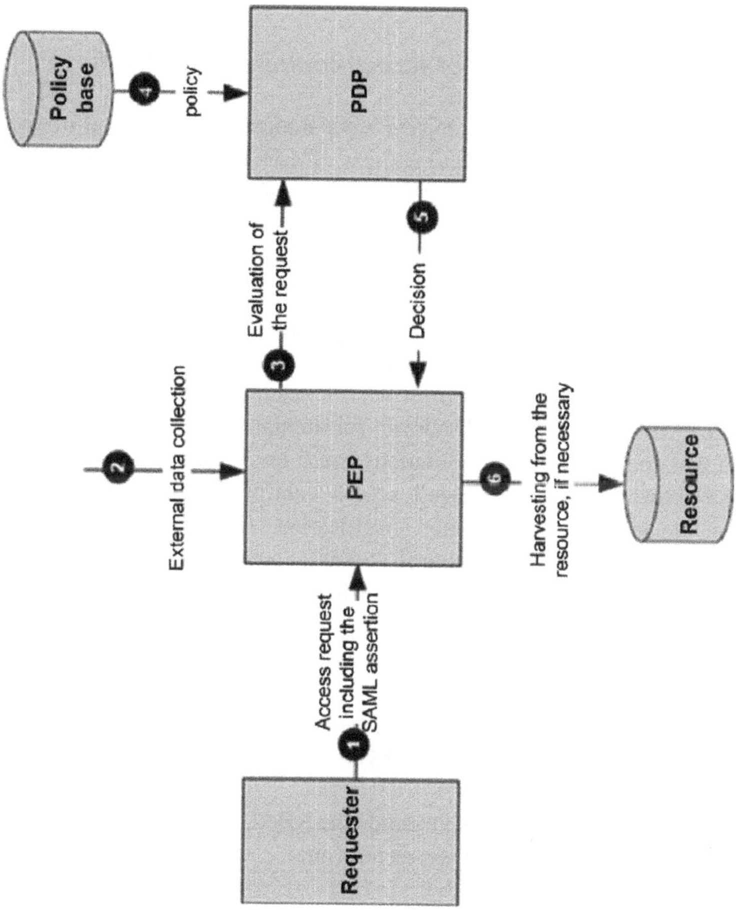

Figure 7.2. *Integrating XACML and SAML (OASIS 2019c)*

7.4. Access control in the IoT

Different European projects and research have presented adaptations to offer the access control service for an emerging IoT environment. We classify this research below, in work on access control models on the one hand (section 7.4.1) and access control systems and technologies on the other hand (section 7.4.2).

7.4.1. *Research on the extension of access control models for the IoT*

A number of European research projects have studied the adaptation of access control security service for the IoT environment. SMARTIE (Secure and SMArter ciTIes data management) (Pokric *et al.* 2015), a European project financed by FP7 (September 2013–December 2016), is focused on access control in the IoT. It aims to develop new mechanisms that establish trust and security in the different layers of the IoT. The project results indicate that access control based on attributes (ABAC) is an adequate solution for a fine-grained access control policy specification in an IoT environment. The identities of IoT service users are not limited to a single attribute, but rely on several attributes (user ID, role, etc.) that form this identity. For these reasons, ABAC type control allows substantial improvements in access authorization and control in the IoT. Solutions based on ABAC make it possible to overcome the drawbacks of centralized access control solutions. Each request requires two stages: an authorization verification (identity monitoring and authentication) and then an access control decision (authorization or prohibition). For each access request, authentication of the IoT service user at their domain is carried out and the access authorization request is obtained from the user. The access authorization request from the user is signed by a trusted authority in the domain. Thus, the user can send the request to IoT devices that verify the signature. If verification of the signature is successful, the information requested is sent to the user (Smartie 2014a, 2014b; Pokric *et al.* 2015).

Furthermore, different research that extends the RBAC model has been proposed for use in the IoT environment and in sensor networks.

For this purpose, we choose the CA-RBAC (Context-Aware-RBAC) model (Garcia-Morchon and Wehrle 2010) and the BTG-RBAC (Break The Glass-RBAC) model (Ferreira *et al.* 2011). The first model, CA-RBAC, makes it possible to provide knowledge of the context and to adapt the security properties as a result. In this model, the decision-making process is divided into three modular context situations: a critical situation, an emergency situation and a normal situation. Based on these situations, the data access privileges will be different (Garcia-Morchon and Wehrle 2010). The second model, BTG-RBAC, makes it possible to gather information needed from final users to define an access control strategy that can be used in emergencies. Indeed, the BTG rule specified by BTG-RBAC enables users to have emergency access to the system when normal authentication does not work or does not work correctly (Ferreira *et al.* 2011).

7.4.2. Research on adapting access control systems and technologies for the IoT

Different European projects and research have proposed different systems and components to ensure access control in the IoT. ARMOUR[1], a European project financed by H2020 (February 2016–February 2018), addressed some security and trust challenges in the IoT. The work carried out as part of this project made it possible to define a set of components that interact to authorize or block secure data requests in an IoT environment. To do this, ARMOUR defines several entities in this environment. In the first place, the PDP is a component that carries access policies enabling an IoT device (a sensor) to be authorized (or not) to carry out an action on a resource (data logging server) by evaluating the access control policies. For example, a "PERMIT" decision coming from the PDP allows the Capability Manager (a server communicating with the PDP) to generate and send a token to the server to publish data on the IoT platform. The Publish/Subscribe Server registers the data and therefore allows execution of the data request to be updated if the token from the sensor received by the Capability Manager permits this action (ARMOUR 2016). Research by Gusmeroli *et al.* (2012) presents an access control system, based on the capacities held by the component requesting access, which businesses or even individuals can use to manage access control for services and information. The proposed mechanism is responsible for delegating rights and a more sophisticated personalization of access control. The proposed access control system is based on several elements:

– the resource: this may be a piece of information for example, the only real constraint is that the resource should be a uniquely identifiable object;

– authorization capacity: this includes information on rights granted (and those that can be delegated), the resource on which these rights can be exercised, the identity of the beneficiary, as well as additional information (e.g. capacity validity period). An authorization capacity is a communicable object and can therefore be provided to the component requesting access using any means of communication. An

1 https://www.armour-project.eu/.

authorization capacity is valid during the specified validity period or until this period is explicitly revoked;

– revocation capacity: this is used to revoke one or more authorization capacities. Like an authorization capacity, a revocation capacity is a transmissible object, with specific rights. It is created to inform the service responsible for managing the resource that specific capacities should not be considered valid. A revocation capacity can revoke a single capacity, a specific capacity and all its descendants, or all the descendants of a specific capacity;

– the service/operation request: this is any request from a service wishing to use the access control method proposed by Gusmeroli *et al.* (2012) accompanied by additional characteristics to reference the authorization capacity to make this request. For example, for a RESTful service, a HTTP GET request on one of the REST resources shown should simply include the capacity to make this request and its proof of ownership to use the access control mechanism proposed;

– the PDP: this is a service responsible for managing validation and decision-making for access requests to resources. It is responsible for validating the capacity of the service request and the concordance also of the request with the access capacity provided, as well as additional access policies, available locally. The result of evaluation of the access request is "authorize" or "refuse";

– the resource manager: this is a service that manages requests and queries for the resource identified. The resource manager should also act as an application point for the PEP by considering the PDP validation results;

– the revocation service: this is responsible for managing capacity revocations. Its objective is both to validate received capacity revocations as well as to update the database of PDP capacities and access policies as a consequence. This means that the PDP should at least ensure that the capacity presented in a service request is acceptable and has not been revoked.

Cruz-Piris *et al.* (2018) have proposed integrating IoT devices into an access control system designed for Web services by modeling some IoT communication elements as resources. This modeling would make it possible to obtain a unified access control schema between heterogenous devices (IoT devices, Internet services, etc.). To achieve this, the authors analyzed the most relevant communication protocols for these types of environments, then proposed a methodology that enables modeling of entities as resources. Then, they protected these resources using access control mechanisms. The proposition was validated by selecting a communication protocol based on the exchange of messages, especially MQTT (Message Queuing Telemetry Transport). MQTT is a lightweight messaging protocol for low-resource IoT devices. It uses publish/subscribe operations to exchange data between clients and the server. The MQTT protocol does not specify any authorization or access control schema. It provides no authorization header in its specification, but the CONNECT message (sent to establish a new connection) contains an optional authorization header, formed of the fields "user name" and "password", that are sent in clear text so long as an encryption system is not applied on the protocol (generally Secure Sockets Layer or SSL). It is possible to use this optional header to transport access control information, such as tokens or tickets, instead of the original user name/password. Thus, the authors of this research profited from this chance to protect resources with the help of UMA (User Managed Access) tokens. UMA is an access management standard based on OAuth[2] which is an authorization standard. The aim of the UMA norm is to enable the owner of a resource to control authorization to share data and other access to this resource between online services that the owner might have or via a third party. The research presented in Cruz-Piris *et al.* (2018) uses the access control process in the UMA specification, which is divided into three phases, to ensure access control in a system based on communications via MQTT between clients and a server. The access control process proposed therefore has three stages: the resource protection stage, the stage of obtaining authorization and finally the stage of executing MQTT actions. The resource protection stage amounts to modeling MQTT subjects (IoT objects, for example) as UMA resources. Just after defining a new MQTT subject, it should be registered to ensure it is protected. The stage of obtaining authorization amounts to the authorization flow, as defined by UMA, for any action abandoning or publishing data. After this stage, the owner of the subject acquires a token, with the authorization needed, which is linked to it, to then be able to transmit or receive data. Finally, during the last stage, the token acquired previously is used to access the resource.

Another piece of research described in Bernabe *et al.* (2016) specifies a new access control system in the IoT. The proposed access control system is flexible and sensitive to trust for the IoT. This system, called TACIoT (Trust Aware Access Control system for the Internet of Things) provides a lightweight authorization mechanism and a new trust model that has been specially designed for IoT environments. TACIoT extends traditional access control systems by considering trust values based on reputation, quality of service, security and social relationship considerations of IoT objects. The access control system adopted in this work is based on a DCapBAC (Distributed Capability-Based Access Control) (Hernandez-Ramos *et al.* 2014). DCapBAC provides a distributed approach that focuses on the

2 https://oauth.net/2/.

application and authorization stage, for which the objects involved can be mutually authenticated and the capacity tokens are exchanged in complete security then validated for a specific access request. To this end, the authorization manager has a PDP, able to take authorization decisions with the help of a policy engine. A "Token Manager" sub-component manages the tokens according to the PDP's decisions. The strategies are defined with the help of the language XACML. XACML also makes it possible to describe the obligations in the policies, which allow the PDP to define directives that should be executed by the PEP before any access is approved. This function is important in the TACIoT proposition, because it is used as a base mechanism to indicate to target devices acting as PEPs that they should authorize access to devices requesting it so long as they satisfy the trust values. Thus, the proposed authorization system enables target devices to consider the trust values of subject devices before allowing them to access their resources. During the token validation process carried out on the target peripheral device, the authorization system deployed on the peripheral device only checks whether the values for the subject or the requester fall below a specific threshold value. If they do, the incoming request is not authorized. Trust values for peripheral devices can be calculated by the "Trust Manager" component, which can be deployed in the target peripheral for peripherals without resource constraints, or in another entity if the peripheral device does not have enough resources. The "Trust Manager" implements a trust model based on fuzzy logic, specially intended for an IoT environment. To calculate the trust, the quality of service dimension considers the number of interactions the object has succeeded in making, the availability of the object, the bandwidth allowed to the out-going and in-coming communications with the object and the delay undergone by the object's data packets. On the other hand, the security dimension, making it possible to evaluate trust, considers not only the encryption and integrity algorithms used, but also the nature of the network to which the object belongs. This trust evaluation model also considers two other dimensions: the object's reputation as well as the object's social relationship, granted following recommendations emanating from other objects or through the object's membership of object groups and of specific clusters.

7.5. Access control framework in the IoT

The IoT environment consists essentially of an interconnection between different infrastructures (Cloud, objects, etc.) with the aim of providing advanced and intelligent services. Thus, these services are provided via the interconnection of IoT objects and the uploading of information from these objects to the Cloud for processing and storage. In this section, we present a proposition for IoT architecture as well as our method of controlling access to objects at IoT gateways, called IoT-MAAC (IoT Multiple Attribute Access Control).

7.5.1. *IoT architecture*

To ensure end-to-end security in an IoT environment, we propose an IoT architecture by implementing security mechanisms at each of the three layers, that is, the sensing layer, the network layer and the Cloud layer (Khalil *et al.* 2020). The lowest layer of this architecture, called the sensing layer, includes the IoT object with constraints in terms of energy consumption as well as LL-Gws (Low-Level Gateways) and HL-Gws (High-Level Gateways) that can be supplied continuously. The second layer, the network layer, includes the network infrastructure making it possible to connect the lower layer with the upper layer of the architecture. This third layer is the Cloud layer where Cloud services are used to process and store the data uploaded by the sensing layer and routed by the network layer. In Figure 7.3, we present our IoT architecture with the security mechanisms that allow the IoT service provider (IoT-SP) to negotiate security needs and requirements with Cloud and network providers as well as IoT service clients via SECSLAs (SECurity Service Level Agreements). Thus, we specify between the IoT-SP and a CSP (Cloud Service Provider) a security service contract called a SECcSLA (SECurity cloud Service Level Agreement) to meet security demands at the Cloud layer. In addition, we implement a security service contract internal to the IoT-SP called a SECgSLA (SECurity gateway Service Level Agreement) to consider the security requirement of the IoT gateways of the sensing layer. As for the network layer of our IoT architecture, no service contract affecting security is agreed with the NSP (Network Service Provider), given that tunnels are put in place between the HL-Gws of the sensing layer and the Cloud to assure all security needs at the level of the network layer. On the other hand, the middle and right-hand tunnels, colored orange in Figure 7.3, are configured by the IoT-SP in the case of a service Cloud service subscription of the IaaS (Infrastructure as a Service) or PaaS (Platform as a Service) types with the CSP. Furthermore, the yellow tunnel on the left in Figure 7.3 is configured by the CSP in the case of a subscription by the IoT-SP to a Cloud service of the SaaS (Software as a Service) type. Access control at the lowest level of the IoT architecture (sensing layer) is defined in the SECgSLA. In this context, our IoT-MAAC access control mechanism is used in this contract and presented in the following sections.

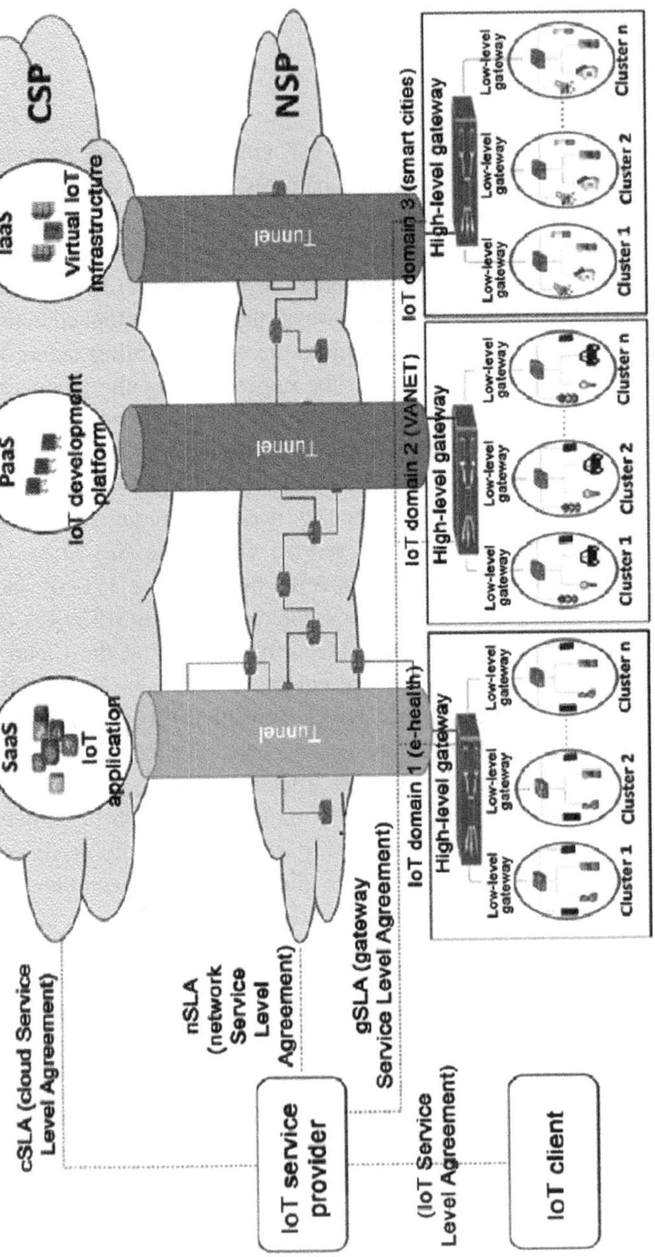

Figure 7.3. *Security architecture for the Internet of Things. For a color version of this figure, see www.iste.co.uk/chalouf/intelligent.zip*

7.5.2. IoT-MAAC access control specification

In our framework, access control involves IoT objects and gateways. In this context, we consider an environment in which different IoT objects are connected to an LL-Gw under the control of a HL-Gw to provide information on an IoT service (see Figure 7.4).

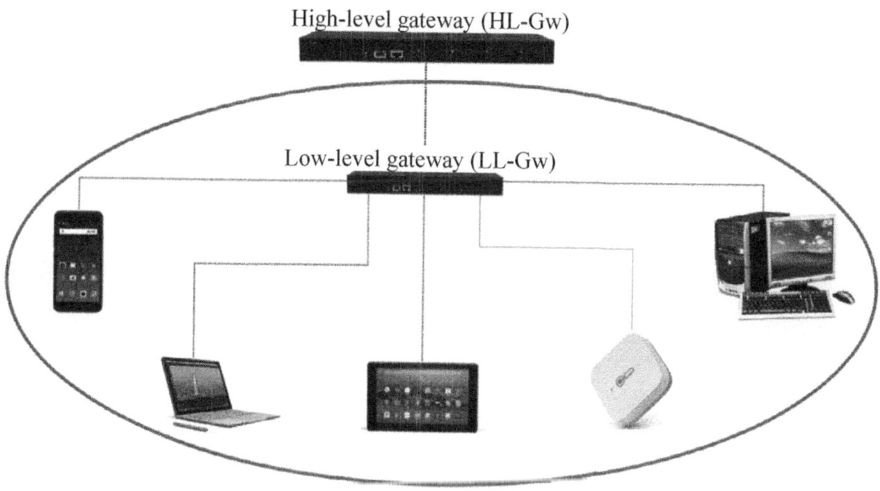

Figure 7.4. *IoT access control environment. For a color version of this figure, see www.iste.co.uk/chalouf/intelligent.zip*

In this IoT access control environment, the LL-Gw can receive critical information from an IoT object that has undergone a physical attack, or indeed a malicious IoT object. Consequently, it is necessary to control the access of IoT objects able to send data about an IoT service to the LL-Gw. Indeed, a corrupted or even malicious IoT object can send erroneous data about a critical event and so endanger good operation of the service as well as the corresponding system.

7.5.2.1. *Access control attributes*

To ensure access control of IoT objects at the LL-Gw, different parameters should be considered to check the identity of and to authorize these objects. Indeed, access control applies to identified and authenticated IoT objects. Thus, in the context of our IoT access control framework at the sensing layer, we propose a new method called IoT-MAAC, allowing access control of IoT objects at the LL-Gw based on four attributes. To do this, we specify as access control attributes for the

IoT-MAAC method: the object identifiers, the object's digital imprint, the object's trust level and the environmental parameters (see Table 7.1).

Object identification is carried out using the identifier of the group of objects (Group Object ID) and the identifier of the object itself (IoT Object ID). These two identifiers are stored in the object and in SECgSLA and SECiSLA type contracts. In addition, the object should transmit its digital imprint including hashing of the characteristics common to the object group to ensure that its integrity is not breached and that it has not been subject to attack. This digital imprint of the object is compared to the object imprint advised in the security contract. As another attribute for our IoT-MAAC access control method, we specify the object trust level, for which we propose an evaluation method based on fuzzy logic and presented in section 7.5.3.3. Finally, with IoT-MAAC we use an attribute relating to environmental information such as the time and date of the access request.

IoT-MAAC attributes	Usage needs
Identifiers (*IoT Object ID, Group Object ID*)	Object identification
Digital imprint	Checking object integrity
Object trust level	Evaluating the objects' trust level
Environment parameters (date and time of the access request)	Evaluating the request environment

Table 7.1. *IoT-MAAC attributes*

7.5.2.2. *Access control architecture based on XACML/SAML*

The access control architecture that we propose (see Figure 7.5) for our IoT framework makes it possible to consider the components of the XACML and SAML standards at the sensing layer of the environment formed by LL-Gws and HL-Gws type gateways, and also the IoT objects (Khalil *et al.* 2020). Below, we detail the components of our architecture making it possible to use the IoT-MAAC method to control access of IoT objects to LL-Gws in the context of offering an IoT service.

A first component of our access control architecture is the HL-Gw that allows us to implement the functions of the PDP and the PAP of the XACML standard as well as a database of access control strategies, a list of SLAs and an attribute authority. In this context, the PAP allows us to safeguard in the list of SLAs the SECgSLAs relating to security of the sensing layer. These SECgSLA type contracts are used to supply not only the attribute authority that carried the information on the objects' trust level, but also the database of strategies relating to the LL-Gws. Furthermore, the PDP uses XACML to interrogate the attribute authority to obtain the trust level corresponding to the IoT object requesting access to the LL-Gw. Details of the method for evaluating the trust level of objects based on fuzzy logic with different security criteria are specified in the following section. The values of these criteria used to calculate the trust level are collected from the SECgSLAs saved in the list of SLAs from the HL-Gw. In the context of our access control architecture, the PDP uses XACML to interrogate the database of strategies to obtain the right strategy to use and consequently the list of policies to apply for access control to the LL-Gw concerned by the request from the IoT object. Indeed, the strategy includes different policies involving a very specific LL-Gw. Each policy concerns a group of objects identified by a Group Object ID and has different rules to apply.

A second component of our access control architecture is the LL-Gw that enables us to implement functions from the policy application point (PEP). This, on the one hand, ensures translation of IoT-MAAC access requests from IoT objects (assertion SAML) into IoT-MAAC XACML requests (XACML authorization requests). On the other hand, it makes it possible to translate IoT-MAAC XACML authorization responses, received from the PDP after the decision has been taken, into IoT-MAAC access control responses. These responses are encapsulated following the communication protocol used in the sensing layer and sent to the IoT objects.

A final component of our access control architecture is the IoT object. This uses an SAML assertion to send an IoT-MAAC access request to the LL-Gw, playing the role of a PEP. This assertion includes different information forming some of the attributes of our IoT-MAAC access method (see Table 7.1). As examples, we list IoT Object ID, Group Object ID and the object's digital imprint. The imprint is calculated using the technique specified in the corresponding SECgSLA (TPM (Trusted Platform Module), IMA (Integrity Measurement Architecture), etc.). The exchanges between the IoT object and the LL-Gw are secured via a symmetrical encryption key established between the two components following the request for access to the LL-Gw. Finally, it should be noted that in our access control architecture, the SP SAML is the HL-Gw that checks the identity of the objects and the IdP SAML is the IoT object itself.

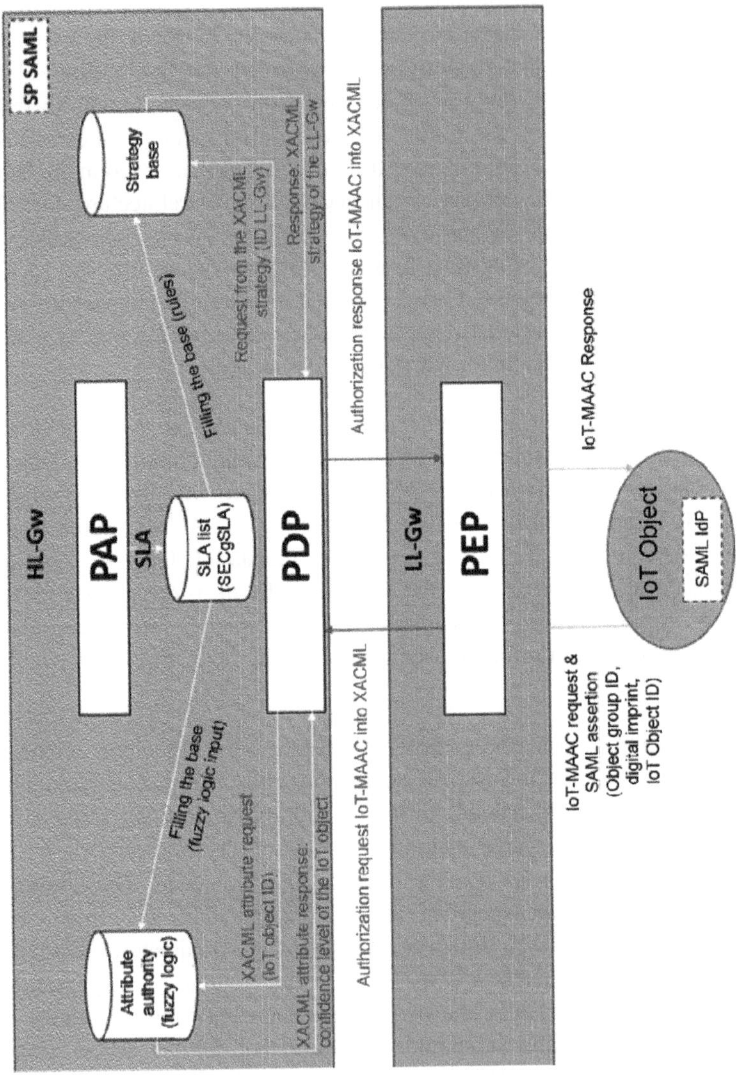

Figure 7.5. *IoT-MAAC-MAAC access control architecture. For a color version of this figure, see www.iste.co.uk/chalouf/intelligent.zip*

7.5.2.3. *Evaluating object trust (TSK fuzzy logic model)*

Fuzzy logic was created by Lotfi Zadeh in 1965 (Dernoncourt 2011). It is a tool to aid decision-making that extends Boolean logic. Fuzzy logic is based on the fuzzy set theory. It represents a set of knowledge base via degrees of condition verification. Thus, each condition will have a degree of verification (degree of exactitude) instead of a true or false type state. This tool thus makes it possible to reproduce human reasoning via a set of rules expressed according to a natural language. Fuzzy logic has been used in different applications. Examples include: image stabilization, pH value control, handover choice in mobile networks, etc. (Scribd 2019). Fuzzy logic uses different important elements making it possible to assure an automated decision-making process. We describe below the components of the fuzzy logic system, that is, the fuzzy sets, the membership functions, linguistic variables, fuzzification, the inference base and defuzzification.

First, fuzzy sub-sets make it possible to implement the notion of weighted membership. Thus, the membership levels of an element of a sub-set will be used. If X is a classic set; a fuzzy sub-set of X, called A, is characterized by a membership function $f_A : X \rightarrow [0,1]$. This is equivalent to the characteristic function of a classic set. In this context, a real value (x) between 0 and 1 is linked to each point x of X. This real value represents the degree of membership of x to A. Consequently, a flexibility of reasoning will be introduced to consider the imprecisions and uncertainties using the notion of degree of membership in verifying a condition.

Second, we have used membership functions. In this context, different forms (sigmoid, hyperbolic tangent, exponential, Gaussian forms, etc.) can be used to characterize these membership functions. These forms will be chosen following advice from an expert, or from statistical studies. The membership function then characterizes one or more sub-sets of an input variable. It is represented in fuzzy logic by the variable $\mu(x)$. Thus, we can represent the membership function of the fuzzy set A of the classic set X by $\mu_A(x)$.

Third, we use linguistic variables. The triplet $(X, V, (x))$ constitute the linguistic variables in a fuzzy sub-set. In this context, X is the variable (classical set of values), V is the variable's value range and (x) is the vocabulary linguistically defining the variable X. This (x) contains the fuzzy sub-sets used to characterize the classic set X. For example, for the temperature, three linguistic variables can be created: weak, average and high, each characterized by a very specific membership function.

Fourth, fuzzification is the process using the membership functions to switch from a real variable to a fuzzy variable. Fuzzification thus makes it possible to determine the degree of membership of a real value to one or more linguistic variables. In this context, different rules formulated with the help of "if" and "then" will be defined by the expert to provide the inference base (no precise directives for establishing the rules of the base). A decision matrix is defined to make it possible to apply and concatenate the rules triggered. To do this, this matrix considers the fuzzy implications based on logical operators (or, and, not).

Finally, defuzzification is the final component in a fuzzy logic system. It is a process making it possible to upload a precise value from fuzzy sets to facilitate decision-making. Several defuzzification methods are available. Thus, we can use the MOM (Mean of Maxima) method or indeed the COG (Center of Gravity) method. The MOM type method calculates the average of the abscissa of the *maxima* of the fuzzy set resulting from the aggregation of the output values. The COG type defuzzification method is based on the abscissa of the center of gravity of the surface of the membership function characterizing the fuzzy set resulting from the aggregation of the output values (Dernoncourt 2011).

Different fuzzy models exist in literature such as the Mamdani model and the Takagi-Sugeno-Kang (TSK) model. Each of these models is specific and can be used in a particular context. The TSK fuzzy logic model relies on a system with r inputs and one output. It is formed of a basis of rules, taking the form: if x_1 is A_1 and x_2 is A_2, and x_r is A_r then $y = f_i(x_1, x_2, \ldots, x_r)$ with $x_1, x_2, \ldots, x_r$ digital input variables of the fuzzy model and f_i is the function that links the inputs to the digital output.

In the TSK model, the evaluation of the output results from the calculation of the weighted average of the triggered rules. In fact, the TSK model makes it possible to calculate for each rule triggered an output value (see equation [7.1]) and a weight for this value. The weight is calculated by applying the arithmetical operator ET to the different degrees of membership of the rule's input parameters. Each membership function in each rule has its degree of membership or truth equal to the value of the y-axis corresponding to the value of the input parameter on the x-axis. The input parameters can trigger one or more rules and, consequently, different weights and output values are calculated:

$$y = f_i(x_1, x_2, \ldots \ldots, x_r) = a_1.x_1 + a_2.x_2 + \cdots \ldots + a_r.x_r \quad [7.1]$$

To evaluate the overall output of our model, the weighted average of the outputs of all the rules triggered is calculated according to equation [7.2] given that N represents the number of rules triggered and w_i represents the weight of an output z_i:

$$Output = \frac{\sum_{i=1}^{N} w_i z_i}{\sum_{i=1}^{N} w_i} \quad [7.2]$$

The TSK model offers flexibility in designing our evaluation system using simple rules (if x_1 is A_1 and x_2 is A_2 then $y = f_i$ $(x_1, x_2, \ldots, x_r)$). Moreover, it is efficient in terms of calculation, because it uses a weighted average to quantify the output, unlike other models that use the COG or MOM methods (Mathworks 2019). This model makes it possible to quantify the output values via digital values and not fuzzy sets to use in the optimization function. The TSK model is used in systems where the output corresponds to a digital level and not to a fuzzy set.

The fuzzy logic model based on TSK which we specify in the context of evaluating one of the attributes of the IoT-MAAC access control method (i.e. the level of trust of IoT objects), considers different input parameters with membership functions specific to each parameter and a set of inference rules that we describe below.

Thus, our TSK type fuzzy logic model uses three input parameters to evaluate the trust level of IoT objects: physical security (DPS: Device Physical Security), the object's security level (DSL: Device Security Level) and the trust accorded to the object's owner (DOT: Device Ownership Trust) (Khalil *et al.* 2019). Different notation systems are proposed to quantify these three input parameters (DPS, DSL and DOT). Each notation system can be personalized with different values, depending on the scenario of IoT use and the characteristics of the corresponding offer of IoT service.

The IoT objects are deployed in diverse physical environments with different security characteristics. In fact, the IoT object can be implemented in environments with heterogenous security levels such as a protected area, a monitored area and an unmonitored area. Furthermore, IoT objects can be protected with a secure loop or can be deployed freely without protection. We specify a notation system to determine the DPS relating to an object. This notation system (see Figure 7.6) makes it possible to determine one note out of a 100 calculated with the help of two sub-parameters, that is the physical environment and the physical security loop.

The object's security level, called DSL, is the second input parameter in our fuzzy logic system for evaluating the trust level of IoT objects. We determine the DSL of the IoT object based on the characteristics of the encryption algorithms and of the hashing used by this object. Similarly, we specify a notation system out of 100 for this DSL parameter whose values result from the sum of the notes allocated to its encryption algorithm and its hashing algorithm (see Figure 7.7).

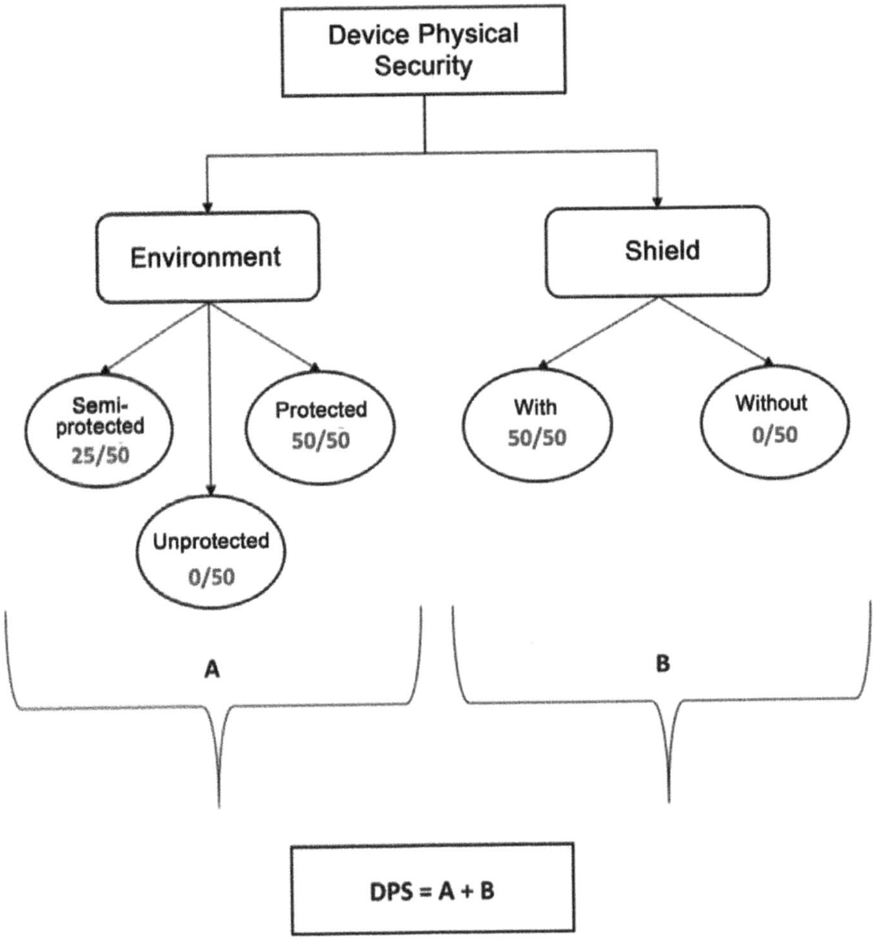

Figure 7.6. *DPS notation system. For a color version of this figure, see www.iste.co.uk/chalouf/intelligent.zip*

The third input parameter in our system for evaluating the trust level of IoT objects is the trust accorded to the object's owner (DOT). This parameter has a direct impact on the overall trust level of IoT objects. Thus, different IoT objects can be used to collect different types of information. These objects can belong to many types of owner, including the IoT-SP, the IoT-C or to third parties. In this context, owners of objects that are untrustworthy can manipulate the IoT objects and consequently modify their features and characteristics, which have an incidence on the trustworthiness and precision of the information collected. To quantify the DOT parameter relating to the trust accorded to the owner of the IoT object, we use a notation system out of 100 depending on the scale described in Figure 7.8.

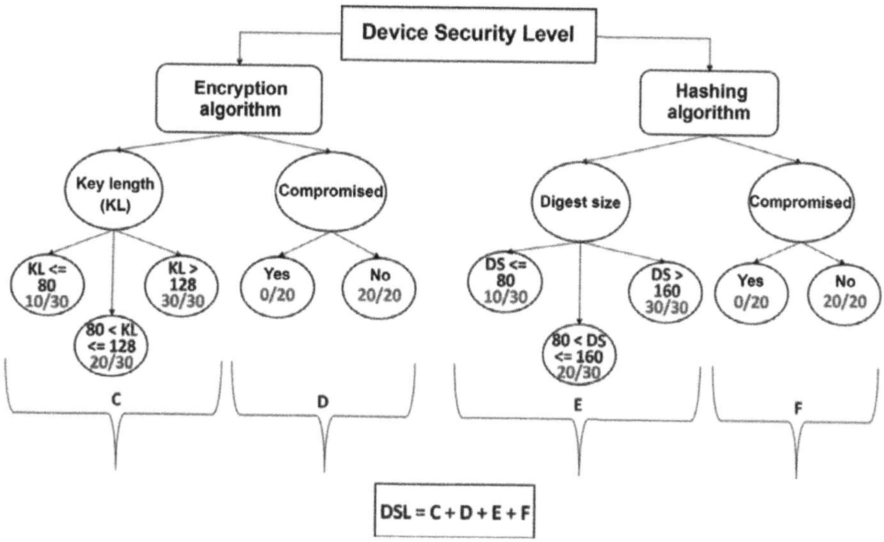

Figure 7.7. *DSL notation system. For a color version of this figure, see www.iste.co.uk/chalouf/intelligent.zip*

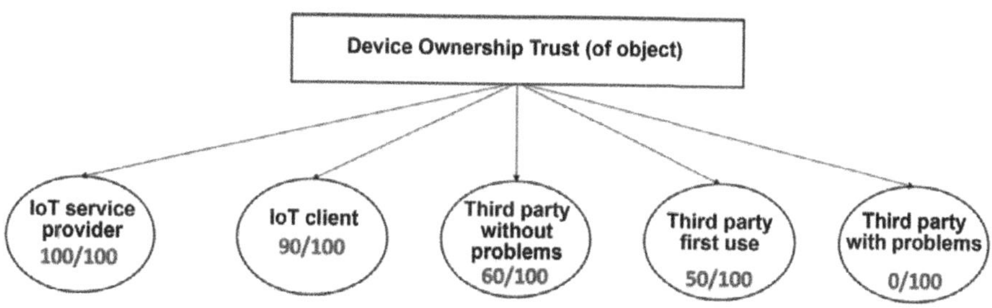

Figure 7.8. *DOT notation system. For a color version of this figure, see www.iste.co.uk/chalouf/intelligent.zip*

In the evaluation system for the trust level of IoT objects, based on fuzzy logic, we use three input parameters (DPS, DSL and DOT). Each of these parameters has a fuzzy set formed of three linguistic variables offering three triangular membership functions (Low, Mid and High) for each parameter. The membership functions that we use for the three input parameters are represented in Figure 7.9.

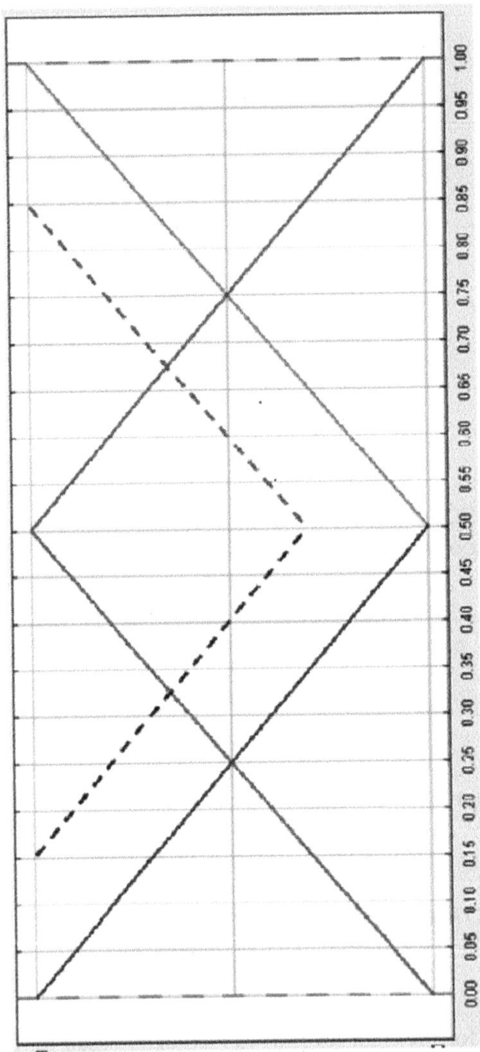

Figure 7.9. *DPS, DSL and DOT membership functions. For a color version of this figure, see www.iste.co.uk/chalouf/intelligent.zip*

The three input parameters of the proposed evaluation system have the same weight, and consequently the same importance in evaluating the trust level of IoT objects. In Table 7.2, we specify the principle behind elaborating the inference base of our TSK model using rules of inference depending on the occurrence of membership function levels (Low, Mid and High). We allocate each rule a well-defined trust output value (from 1 to 10, with 10 the best possible trust level).

Occurrence	Trust output value
3 Low	1
2 Low and 1 Mid	2
2 Low and 1 High	3
2 Mid and 1 Low	4
3 Mid	5
1 Low, 1 Mid and 1 High	6
2 Mid and 1 High	7
2 High and 1 Low	8
2 High and 1 Mid	9
3 High	10

Table 7.2. *Principle of the inference base of the TSK model for evaluating the trust of IoT objects*

7.5.2.4. *Decision-making algorithm*

The IoT-MAAC access control method uses a First Applicable (this means that the first rule verified is applied) type decision-making algorithm. Below, we describe how the decision-making algorithm illustrated in Figure 7.10 operates.

The IoT-MAAC decision-making algorithm implemented at the level of the PDP in the HL-Gw relates to access to an IoT object to an LL-Gw. It makes it possible to determine first the strategy linked to this LL-Gw as well as the number of policies forming this strategy. Then, the algorithm allows us to browse these policies to determine, thanks to the value of the Group Object ID, the one that relates to the IoT object requesting access. The policy linked to this object, like each of the other policies in the strategy, still includes two rules that will be tested successively. Thus, the decision-making algorithm begins by testing in order the different attributes of the first rule. To do this, the algorithm tests the number of attributes in the rule. If a single attribute exists in the rule in question, then a decision to refuse access is taken because the attributes of the request received do not meet the attributes defined in the first rule of the policy. If several attributes exist, the first rule is tested and consequently the different attributes will be evaluated by respecting the following order: test the inclusion of the IoT Object ID in the Group Object ID list of objects, test the object's digital imprint (fingerprint), test the object's trust level and test the access time (Request Access Time). Thus, the decision-making algorithm begins by verifying the attributes in this order. If an attribute of the object does not correspond to the attribute in the rule, then the algorithm allows us to switch directly to the following rule (the second rule corresponding to refusal of access). On the other hand, if all the attributes of the rule are met, then authorization for access is accorded to the IoT object. It should be noted that the IoT-MAAC decision-making algorithm predicts, in the case where no policy corresponds to the Group Object ID uploaded by the object, that a non-applicability decision is uploaded, which corresponds to an implicit access refusal. This non-applicability decision results from the fact that the strategy of the LL-Gw does not include a policy and consequently rules on the Group Object ID of the object requesting access.

```
Decision Making Algorithm [LL-Gw_ID]

Strategy ← Retrieve_Strategy_from_Strategy_Directory (LL-Gw_ID)
Number_Policies ← Retrieve_Number_Policies (Strategy);
for i ∈ (1, Number_Policies) do
        if ( Group_Object_ID (IoT Object) = Group_Object_ID (Policy(i)) ) then
                Number_Rules ← 2;
                for i ∈ (1, Number_Rules) do
                        if ( Attributes (Rule (i) = 1) ) then
                                return Deny;
                        else
                            if [( (IoT Object ID) ∈ (List_Object [Group Object ID]) )
                              &( Fingerprint (IoT Object) = Fingerprint (Rule(i)) )
                                & ( Trust_Level (IoT Object) >= Trust_Level (Rule(i)) )
                                  & ( Request_Access_Time (Start) >= Start_Time(Rule(i)))
                                    & ( Request_Access_Time(End) < End_Time(Rule(i)))] then
                                        return Permit;
                            end if
                        end if
                end for
        end if
end for
return NotApplicable
```

Figure 7.10. *IoT-MAAC decision-making algorithm. For a color version of this figure, see www.iste.co.uk/chalouf/intelligent.zip*

7.5.2.5. *Usage scenario*

In the context of the IoT-MAAC access control method, we specify different exchanges where the format respects SAML and XACML standards. Thus, the IoT object uses an SAML assertion (see Figure 7.11) to communicate with, among others, one of the attributes (IoT Object ID, fingerprint, etc.) of IoT-MAAC control of access to the LL-Gw.

```xml
<saml:Assertion xmlns:saml="urn:oasis:names:tc:SAML:2.0:assertion"
  Version="2.0"
  IssueInstant="2019 02 12T15:26:12Z">
    <saml:Issuer Format="urn:oasis:names:SAML:2.0:nameid-format:entity">
      IoT Object 1
    </saml:Issuer>
    <saml:Subject>
      <saml:NameIdentifier>IoT Object 1</saml:NameIdentifier>
    </saml:Subject>
    <saml:Conditions NotBefore="2019-02-12T15:26:12Z" NotOnOrAfter="2019-02-12T15:28:12Z"/>
    <saml:AttributeStatement>
      <saml:Attribute Name="Fingerprint">
        <saml:AttributeValue>$oàikj^plnrusdn=çkdjd21587uyrzertzvssrh</saml:AttributeValue>
      </saml:Attribute>
      <saml:Attribute Name="IoT_Object_ID">
        <saml:AttributeValue>00:50:56:C0:00:01</saml:AttributeValue>
      </saml:Attribute>
      <saml:Attribute Name="Group_Object_ID">
        <saml:AttributeValue>1</saml:AttributeValue>
      </saml:Attribute>
    </saml:AttributeStatement>
</saml:Assertion>
```

Figure 7.11. *SAML assertion of the IoT object. For a color version of this figure, see www.iste.co.uk/chalouf/intelligent.zip*

```
<Request xmlns="urn:oasis:names:tc:xacml:2.0:context:schema:os" xmlns:xsi="http://www.w3.org/2001/XMLSchema-instance">
  <Subject>
    <Attribute AttributeId="urn:oasis:names:tc:xacml:1.0:subject:subject-id" DataType="http://www.w3.org/2001/XMLSchema#string">
      <AttributeValue>IoT Object 1</AttributeValue>
    </Attribute>
    <Attribute AttributeId="Fingerprint" DataType="http://www.w3.org/2001/XMLSchema#string">
      <AttributeValue>$oàikj^plnrusdn=çkdjd21587</AttributeValue>
    </Attribute>
    <Attribute AttributeId="IoT_Object_ID" DataType="http://www.w3.org/2001/XMLSchema#string">
      <AttributeValue>11:12:13:14:15:16</AttributeValue>
    </Attribute>
    <Attribute AttributeId="Group_Object_ID" DataType="http://www.w3.org/2001/XMLSchema#string">
      <AttributeValue>1</AttributeValue>
    </Attribute>
  </Subject>
  <Resource>
    <Attribute AttributeId="urn:oasis:names:tc:xacml:1.0:resource:resource-id" DataType="http://www.w3.org/2001/XMLSchema#string">
      <AttributeValue>11:15:16:18:17:16</AttributeValue>
    </Attribute>
  </Resource>
  <Action>
    <Attribute AttributeId="urn:oasis:names:tc:xacml:1.0:action:action-id" DataType="http://www.w3.org/2001/XMLSchema#string">
    <AttributeValue>read</AttributeValue>
    </Attribute>
    <Attribute AttributeId="urn:oasis:names:tc:xacml:1.0:action:action-id" DataType="http://www.w3.org/2001/XMLSchema#string">
    <AttributeValue>write</AttributeValue>
    </Attribute>
  </Action>
  <Environment>
      <Attribute AttributeId="NotBefore" DataType="http://www.w3.org/2001/XMLSchema#datetime">
    <AttributeValue>2019-02-12T15:26:12Z</AttributeValue>
    </Attribute>
    <Attribute AttributeId="NotAfter" DataType="http://www.w3.org/2001/XMLSchema#datetime">
    <AttributeValue>2019-02-12T15:26:12Z</AttributeValue>
    </Attribute>
</Request>
```

Figure 7.12. *XACML authorization request from the LL-Gw in IoT-MAAC. For a color version of this figure, see www.iste.co.uk/chalouf/intelligent.zip*

In Figure 7.12, we present an example of an SAML assertion sent by an IoT object that contains information on the validity of this assertion (Not Before, etc.) as well as the fingerprint type access control attribute, and also the object identification information such as *IoT_Object_ID* and *Group_Object_ID*.

In the context of IoT-MAAC access control, this SAML assertion is translated by the PEP (LL-Gw) into an IoT-MAAC XACML authorization request (see Figure 7.12) to send to the PDP (HL-Gw) which should take a decision by running the IoT-MAAC decision-making algorithm.

The XACML request includes not only information harvested from the SAML assertion, but also information on the requested resource (identifier of the LL-Gw) and access rights to the resource (reading, writing, modification, etc.) specified by the resource itself. In fact, the LL-Gw specifies in the rights request for access to it by the object in question, so that these rights can be studied by the PDP at the HL-Gw. Furthermore, a time interval is specified in this XACML request to define the access time requested. Figure 7.12 shows an example of a XACML request sent by the LL-Gw to the HL-Gw.

```
<Response xmlns="urn:oasis:names:tc:xacml:2.0:context:schema:os" xmlns:xsi="http://www.w3.org/2001/XMLSchema-instance">
    <Result>
        <Decision>Permit</Decision>
        <Status>
            <StatusCode Value="urn:oasis:names:tc:xacml:1.0:status:ok"/>
        </Status>
    </Result>
</Response>
```

Figure 7.13. *XACML response from the HL-Gw in IoT-MAAC. For a color version of this figure, see www.iste.co.uk/chalouf/intelligent.zip*

After the decision is taken by the HL-Gw playing the role of PDP in the IoT-MAAC architecture, an XACML response is sent to the LL-Gw playing the role of PEP. In Figure 7.13, we present an example of a positive IoT-MAAC XACML authorization response "permit". This response will then be translated by the PEP into a final IoT-MAAC access response that can be understood by the object following the communication technology used between this object and the LL-Gw.

7.6. Conclusion

It is necessary to adapt security mechanisms by considering the characteristics of the IoT environment in order to provide an adequate level of security for users of IoT services. One of the essential security services in an IoT environment is access control. This vital service makes it possible to control access to the resources in this environment and to provide reliable IoT services. Through this chapter, we have presented the specifics of security services, including access control in an IoT environment characterized by limited resources. Then, we studied the mechanisms and also the techniques and technologies used for access control via several pieces of research allowing for the adaptation of existing access control solutions for the IoT environment. Finally, we have presented a security framework for the IoT including an architecture proposition and an access control method called IoT-MAAC and based on XACML and SAML standards. This access control method is adapted to an IoT environment. It uses different attributes including the trust level of the IoT objects, which is quantified via a fuzzy logic model considering input parameters specific to IoT objects.

7.7. References

Abdemeziem, M. (2016). Data confidentiality in the Internet of Things. PhD Thesis, University of Lorraine, Nancy.

ARMOUR (2016). ARMOUR experiments and requirements version 1.0. Document, ARMOUR.

Atlam, H.F., Alassafi, M.O., Alenezi, A., Walters, R., Wills, G. (2018). XACML for building access control policies in Internet of Things. In *3rd International Conference on Internet of Things, Big Data and Security*. INSTICC, Funchal.

Balte, A., Kashid, A., Patil, B. (2015). Security issues in Internet of Things (IoT): A survey. *International Journal of Advanced Research in Computer Science and Software Engineering*, 5(4).

Bernabe, J.B., Hernandez-Ramos, J., Skarmeta, A.F. (2016). TACIoT: Multidimensional trust-aware access control system for the Internet of Things. *Soft Computing*, 20(5), 1763–1779.

Cruz-Piris, L., Rivera, D., Marsa-Maestre, I., De La Hoz, E., Velasco, J.R. (2018). Access control mechanism for IoT environments based on modelling communication procedures as resources. *Sensors*, 18, 917.

Dernoncourt, F. (2011). Introduction à la logique floue [Online]. Available at: https://www.researchgate.net/publication/267041176_Introduction_a_La_Logique_Floue.

Ferreira, A., Correira, R., Monterio, H., Brito, M., Antunes, L. (2011). Usable access control policy and model for healthcare. In *24th International Symposium on Computer-Based Medical Systems*. IEEE, Bristol.

Garcia-Morchon, O. and Wehrle, K. (2010). Modular context-aware access control for medical sensor networks. In *ACM Symposium on Access Control Models and Technologies*. ACM, Pittsburgh.

Gusmeroli, S., Piccione, S., Rotondi, D. (2012). IoT access control issues: A capability based approach. In *6th International Conference on Innovative Mobile and Internet Services in Ubiquitous Computing*. IEEE, Palermo.

Hernandez-Ramos, J.L., Jara, A.J., Marin, L., Skarmeta, A.F. (2014). DCap-BAC: Embedding authorization logic into smart things through ECC optimizations. *International Journal of Computer Mathematics*, 93(2), 345–366.

ITU-T (1991). Architecture de sécurité pour l'interconnexion en systèmes ouverts d'applications du CCITT. Document, ITU-T.

Joshi, N. (2017). Authentication and device identification in IoT security. Document, Allerin, Navi Mumbai.

Khalil, A., Mbarek, N., Togni, O. (2019). Fuzzy logic based security trust evaluation for IoT environments. In *16th International Conference on Computer Systems and Applications*. IEEE/ACS, Abu Dhabi.

Khalil, A., Mbarek, N., Togni, O. (2020). IoT-MAAC: Multiple attribute access control for IoT environments. In *IEEE 17th Annual Consumer Communications & Networking Conference (CCNC)*. IEEE, Las Vegas.

Li, S. (2017). IoT node authentication. In *Securing the Internet of Things*, Li, S., Xu, L.D. (eds). Elsevier, Cambridge.

Mathworks (2019). Mamdani and Sugeno fuzzy inference systems [Online]. Available at: https://fr.mathworks.com/help/fuzzy/types-of-fuzzy-inference-systems.html;jsessionid=3d8750953818e96ed6ee074c149b [Accessed 19 September 2019].

Maw, H., Xiao, H., Christiansin, B., Malcolm, J. (2014). A survey of access control models in wireless sensor networks. *Journal Sensor and Actuator Networks*, 3(2), 150–180.

Mosenia, A. and Jha, N. (2017). A comprehensive study of security of Internet of Things. *IEEE Transactions on Emerging Topics in Computing*, 5(4), 586–602.

OASIS (2019a). Standards [Online]. Available at: https://www.oasis-open.org/standards [Accessed 14 February 2019].

OASIS (2019b). OASIS eXtensible Access Control Markup Language (XACML) TC [Online]. Available at: https://www.oasis-open.org/committees/tc_home.php?wg_abbrev=xacml [Accessed 14 February 2019].

OASIS (2019c). SAML V2.0 technical overview [Online]. Available at: http://docs.oasis-open.org/security/saml/Post2.0/sstc-saml-tech-overview-2.0-cd-02.pdf [Accessed 19 September 2019].

Pokric, B., Krimmling, J., Soika-Piotrowska, A., Dyka, Z., Lagendorfer, P., Bohli, J., Marin-Perez, R. (2015). Node security, policies, and automatic configuration version 1.0. Document, SMARTIE.

SAML (2019). SAML specifications [Online]. Available at: http://saml.xml.org/saml-specifications [Accessed 23 April 2019].

Scribd (2019). Motivations of fuzzy logic [Online]. Available at: https://www.scribd.com/document/172969317/Motivations-of-Fuzzy-Logic-doc [Accessed 4 April 2019].

SMARTIE (2014a). IoT information access and privacy preservation. Document, SMARTIE.

SMARTIE (2014b). Components for secure information gathering and storage. Document, SMARTIE.

XACML (2019). eXtensible Access Control Markup Language (XACML) version 3.0 [Online]. Available at: http://docs.oasis-open.org/xacml/3.0/xacml-3.0-core-spec-os-en.html [Accessed 23 April 2019].

8

The Contributions of Biometrics and Artificial Intelligence in Securing the IoT

Amal SAMMOUD[1], Omessaad HAMDI[2], Mohamed-Aymen CHALOUF[3] and Nicolas MONTAVONT[4]

[1] *SYSCOM, ENIT, University of Tunis El Manar, Tunisia*

[2] *IEEE, IMT Atlantique, Rennes, France*

[3] *IRISA, University of Rennes 1, Lannion, France*

[4] *IRISA, IMT Atlantique, Rennes, France*

8.1. Introduction

The Internet of Things (IoT) appears today in many domains such as surveillance, health, agriculture, etc. This new technology is based on connected physical objects able to communicate with one another, and also with other devices. The services offered thanks to the IoT are accessible to a user anywhere at any time (Dhillon and Kalra 2017). Security and protection of user privacy are a priority for this new technology to be fully accepted. To meet these security imperatives, biometrics as well as artificial intelligence could make it possible to provide fairly robust security solutions, well-adapted to the IoT.

In this chapter, we focus on the contributions of biometrics and artificial intelligence in optimizing the security of some IoT applications. In particular, we will study the most reliable security approaches, which are based on using biometric methods and artificial intelligence techniques to ensure safe authentication in the IoT.

This chapter is structured as follows. Section 8.2 presents the security requirements of the IoT and connected objects. In section 8.3, we introduce the principles of authentication based on biometrics. In section 8.4, we present multifactor authentication techniques based on biometrics, especially for the IoT. Section 8.5 details authentication techniques based on biometrics and machine learning. Indeed, in this section we present security approaches using artificial intelligence techniques, such as machine learning, to ensure authentication based on a biometric factor (ECG: electrocardiogram). Section 8.6 describes the challenges and limits of authentication based on biometrics. Section 8.7 concludes the chapter.

8.2. Security and privacy in the IoT

The IoT, which integrates a variety of connected objects to provide services to users, should ensure security services and protect user privacy by ensuring protection against attacks such as identity theft, Denial of Service (DoS), passive listening and unauthorized access to personal and sensitive data (Xiao *et al.* 2018). With the development of new technologies, the IoT is now faced with intelligent attacks. To resist this type of attack, IoT devices should choose a defense policy and set key security parameters in the protocols used (Xiao *et al.* 2018). This task can prove complicated, since connected objects often have restricted resources (in terms of calculation, memory, bandwidth and battery), which do not allow them to run the necessary processing correctly in a reasonable time and without compromising lifespan (Xiao *et al.* 2018). To do this, most security solutions proposed for the IoT demand optimization of the computational load and communication to be used in connected objects.

Security in the IoT happens mainly by protecting connected objects by ensuring mechanisms such as authentication, access control and detection of malware:

– authentication helps objects to distinguish the source node and to protect against identity theft type attacks;

– access control prevents unauthorized users from accessing information and data stored on the objects;

– malware detection makes it possible to protect objects from malicious software such as viruses, worms and Trojans, and so to avoid the consequences that may occur, such as data leaks, exhaustion of the power supply and the degradation of performances.

In this chapter, we focus mainly on authentication protocols intended for the IoT. In this fairly vulnerable environment, authentication mechanisms should consider the vulnerabilities of connected objects by guaranteeing the following security requirements (Lee *et al.* 2020):

– user anonymity: ensuring that it is impossible for an attacker to discover the real identity of the user;

– untraceability: preventing an attacker from being able to follow the user's activities;

– mutual authentication: guaranteeing mutual authentication between different communicating entities to verify the legitimacy of each;

– resistance to diverse attacks: ensuring high-level protection against known attacks such as replay attacks, identity theft attacks and insider attacks.

Among the robust authentication methods that are at the same time adapted to the IoT environment, we find authentication based on biometric methods. This type of authentication can be used in various IoT domains such as health and smart homes. With the development of artificial intelligence techniques such as machine learning, we today find authentication mechanisms based on biometric methods that are processed using machine learning techniques.

8.3. Authentication based on biometrics

8.3.1. *Biometrics*

The word biometric comes from the Greek words *bios* (life) and *metrikos* (measurement). Biometric authentication is a technique that calls on the statistical analysis of biological characteristics. It consists of using physical characteristics to identify a person. These characteristics should be reliable, unfalsifiable, universal, unique for each individual, measurable and recordable.

Biometrics is therefore used in many applications entering a location, accessing work station, etc. It is even beginning to replace classical methods, which depend on what the user knows (password, PIN, etc.) or possesses (key, card, etc.), that can be discovered, modified or stolen.

8.3.2. *Biometric techniques*

In the main, there are two types of biometric techniques that are used for identification applications (see Figure 8.1):

– physiological techniques, based on identifying physical characteristics that are unique. This category includes the iris, the retina, handshape, fingerprints, facial features, etc.;

– behavioral techniques, based on analyzing people's behavior. This category includes voice recognition, signature dynamic, key stroke dynamic, etc.

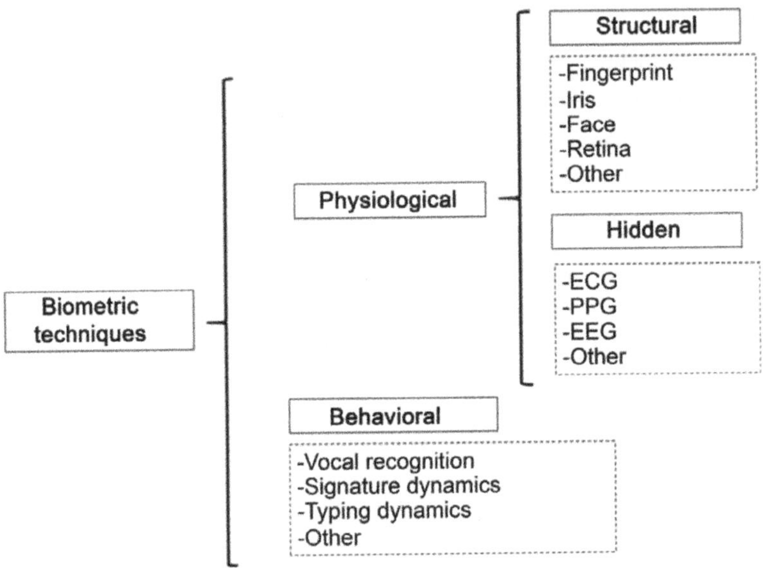

Figure 8.1. *Biometric techniques (Cabra* et al. *2018)*

There is another technique, called biological or anatomical, which is the study of biological traces such as DNA, blood, saliva, urine and odor. Today, this technique is not used in authentication.

8.3.3. *The different properties of biometrics*

Biometric characteristics making it possible to verify the identity of an individual should have the following properties (Ometov *et al.* 2018):

– universality: the biometric should be present in each person;

– uniqueness: the biometric should differ sufficiently between one person and another;

– stability: the properties of the biometric should remain invariable throughout the person's life;

– collectability: the biometric should be easy to measure in an individual;

– performance: the biometric should be sufficiently robust, reliable and easy to analyze when it is used to verify a person's identity;

– acceptability: use of the biometric should be socially acceptable;

– identity theft: the biometric should not be easy for an attacker to steal.

Biometric characteristics possess these properties to different degrees as Table 8.1 shows. None of these different biometric characteristics is ideal, but each of them can be taken and adapted to particular applications.

Biometrics	A	B	C	D	E	F
Voice	A	L	A	L	H	H
Face	H	L	A	H	H	A
Fingerprint	A	H	A	H	A	H
Handshape	A	A	A	A	A	A
Vein	A	A	A	A	A	A
Behavior	H	H	L	L	L	L
OCS	NA	L	L	L	L	A
ECG	L	H	L	A	A	L
EEG	L	H	L	A	L	L
DNA	H	H	L	H	L	L

H: high; A: average; L: low; NA: not available.

Table 8.1. *Comparison of biometric characteristics according to the following properties: universality (A), uniqueness (B), collectability (C), performance (D), acceptability (E) and identity theft (F) (Ometov* et al. *2018)*

8.3.4. *Operating a biometric system*

A biometric system makes it possible to sense biometric information, analyze it and extract characteristic elements from it and compare these with others memorized in a database.

Using a biometric system requires two stages:

– an enrollment stage (also called a learning stage), which consists of sensing, modeling (constructing a model of the individual) and storing biometric data in the database;

– a reconnaissance stage, which consists of comparing the new sensing with the assumed model of the individual.

The reconnaissance will differ depending on the mode of identification or verification:

– identification mode: to guess the identity of the person, the system answers the question "Who am I?". In this mode, the system compares the signal measured with the models stored in the database. We often speak of a closed problem (each person using the system has a model in the database);

– verification mode: to verify the person's identity, the system answers the question "am I the person I claim to be?". The person proposes an identity to the system and the system verifies that the identity of the person is really the one proposed. We call it an open problem (we suppose that a person who has no model in the database may seek to be recognized).

8.3.5. *System performances*

It is impossible to obtain a "signature" signal created during verification that is similar to a "signature" signal created at the moment of learning. It is impossible to reproduce original elements identically (an image, an ECG signal, a sound, etc.).

The performances of biometric authentication systems are measured by:

– FRR (False Rejection Rate): this is the proportion of legitimate users wrongly rejected;

– FAR (False Acceptance Rate): this is the proportion of users accepted/identified erroneously.

8.4. Multifactor authentication techniques based on biometrics

A robust authentication protocol plays a very important role in guaranteeing the security and privacy of users of the IoT. Today, security protocols based on biometrics are much used to ensure authentication or verification of identity in environments that demand a fairly high level of security such as e-health. One of the most significant advantages of technology based on biometrics is that the characteristics used as identifiers cannot be lost or replaced, and it is very difficult to falsify them. Unlike other authentication technologies, biometric characteristics cannot be forgotten, nor stolen. For example, passwords are easily forgotten, and smart cards can be lost or stolen (Hamidi 2019). On the contrary, using different authentication technologies at the same time (password, smart card and biometric) can only strengthen the level of security.

8.4.1. *Multifactor authentication*

In general, authentication is the procedure that enables a system (e.g. an IoT device) to verify whether an entity (e.g. another IoT device or even a user via their terminal) is authorized or not to access resources controlled by this system. Authentication methods are based on authentication factors such as the password, smart card and/or biometrics (Figure 8.2). There are two authentication categories:

– simple authentication, which requires only a single authentication factor (e.g. a password and badge);

– strong authentication (also called multifactor authentication), which calls on many factors simultaneously (at least two).

In general, a reliable authentication protocol is based on three security factors (Table 8.2):

– a memory factor (what you know): information that is known only by the user (e.g. passwords and PIN codes);

– a material factor (what you have): a physical key that belongs only to the user (e.g. smart card or mobile phone);

– a bodily factor (who you are, biometrics): biometric data are unique for each individual (e.g. voice, fingerprints and ECG).

Authentication factor	Advantages	Drawbacks
Material factor (token, card, etc.)	A new token can be supplied in case of loss	Can be stolen A fake token can be sent Can be shared The same person can be registered with different identities
Memory factor (password, PIN, etc.)	Simple and economic method Can easily be replaced by a new one	Can be guessed Good passwords are difficult to remember Can be shared
Bodily factor (biometrics)	Cannot be guessed, forgotten, stolen or shared Offers a higher level of security compared to others	Impossible to replace it

Table 8.2. *Description of authentication factors*

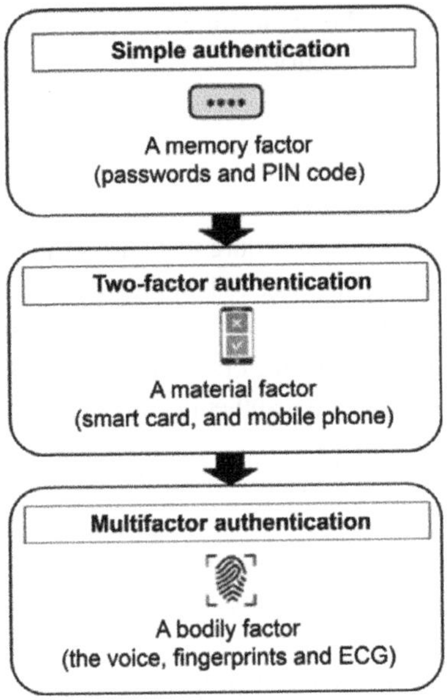

Figure 8.2. *The evolution of authentication methods from simple authentication to strong authentication (Ometov and Bezzateev 2017). For a color version of this figure, see www.iste.co.uk/chalouf/intelligent.zip*

In this context, many researchers have developed authentication protocols based on three factors (a password, a smart card and biometrics) to ensure security of communications in an IoT network (Dhillon and Kalra 2017; Ali *et al.* 2018; Ahmed and Ahmed 2019; Li *et al.* 2019; Sammoud *et al.* 2020c).

In fact, authentication protocols based on a single factor (e.g. a password) do not provide reliable authentication (Sammoud *et al.* 2020c). We also note that the majority of authentication protocols that are based on two factors, memory and material, have not managed to guarantee good resistance to some significant attacks such as identity guessing attacks, insider attacks and password guessing attacks (Sammoud *et al.* 2020c). To do this, authentication protocols based on three factors have been proposed. They have the objective of improving security of authentication in the IoT.

8.4.2. *Examples of multifactor authentication approaches for securing the IoT*

Several pieces of research focus on the security of the IoT via multifactor authentication approaches. In this context, in 2017, Dhillon and Kalra proposed a multifactor authentication approach based on biometrics, dedicated to services offered by the IoT. Formal and informal analysis of this proposition demonstrates its robustness in terms of security. In 2019, Ahmed and Ahmed proposed a multifactor authentication approach to securing the IoT. The authors show that the proposed approach is robust against many types of attack and also guarantees the integrity, authenticity and availability of detected data for legitimate devices in the IoT.

The IoT has contributed to the dramatic evolution of services offered in the healthcare sector. It is very important to guarantee the protection of sensitive data exchanged in this domain. In 2018, Ali *et al.* proposed an authentication protocol based on three factors for securing the IoT in the context of e-health systems. They have demonstrated the robustness of the security of their protocol using BAN (Burrows-Abadi-Needham) logic and security analysis with the formal verification tool, AVISPA. In 2019, Li *et al.* proposed a multifactor authentication protocol for the wireless medical sensor network. The security of this protocol has been demonstrated by the formal verification tool ProVerif. Recently, Sammoud *et al.* (2020c) proposed an optimal protocol for multifactor authentication and generation of session keys to secure direct communications between healthcare professionals and the patient's medical sensors. Using analysis of formal and informal security, they have demonstrated that their protocol is robust enough against the most well-known attacks such as password attacks, replay attacks and attacks on the gateway or sensor, or attacks that involve stealing the user's identity. In addition, the authors of Sammoud *et al.* (2020c) have shown that their protocol is more efficient in terms of the resources required (calculation, memory and energy) than the protocols proposed by Ali *et al.* (2018) and Li *et al.* (2019).

8.4.3. *Presentation of the approach of Sammoud et al. (2020c)*

Sammoud *et al.* (2020c) have proposed a multifactor authentication protocol with generation of a session key to secure a direct communication between medical personnel and a patient's network of wireless body sensors (WMSN: Wireless Medical Sensors Network). In this type of communication, with a terminal (wired, wireless or cellular) and Internet access, a doctor (or more generally, a healthcare professional) can access a medical sensor situated in the patient's WMSN network via a gateway. This access is generally achieved in the following way. First of all, a smartcard is delivered to the healthcare professional by the administration system. This smart card is then used to verify the legitimacy of the healthcare professional and to send a connection message to the gateway via a public channel (e.g. the Internet). Then, via the gateway, the healthcare professional and the medical sensor agree on a session key to secure later communications.

The authors of Sammoud *et al.* (2020c) have shown that the proposed protocol ensures a robust protection against known attacks while still ensuring low consumption of resources (CPU, memory and energy). Indeed, formal and informal security analysis of the protocol has demonstrated its robustness against password guessing attacks, user identity guessing attacks, gateway impersonation attack, sensor impersonation attack, insider attacks and replay attacks, as well as maintaining user untraceability and protection of the symmetric key generated, which will be used later to secure communications between the healthcare professional and the medical sensor against illicit disclosure. In addition, by comparing the protocol with other multifactor authentication protocols, the authors of Sammoud *et al.* (2020c) have demonstrated that their proposition has the advantage of resisting a large number of attacks while still minimizing consumption of resources (calculation, memory and energy).

Like most work in this domain (defining multifactor authentication), this approach has five phases: setup, healthcare professional registration, login, authenticating and password change phase.

8.4.3.1. *Notation*

Table 8.3 introduces the notations that will be used to describe the approach of Sammoud *et al.* (2020c).

Notation	Explanation
U_i	Healthcare professional (user) i
GW	Gateway
SN_j	Sensor node j
ID_i	User identifier U_i
PW_i	User password U_i
B_i	Biometric characteristic of the user U_i
ID_g	Gateway identifier GW
SID_j	Sensor node identifier SN_j
SK	Symmetric key, generated to secure communication between U_i and SN_j
Gen, Rep	Probabilistic generation procedure and deterministic reproduction procedure of the fuzzy extractor
$h(\cdot)$	Hashing function
⍰	XOR Operator
$\|\|$	Concatenation operator

Table 8.3. *Notation table*

8.4.3.2. *Setup phase*

First of all, the administration system *SA* selects an identity SID_j for each sensor node and calculates $K_{snj} = h\ (SID_j\ //\ K)$, where *K* represents a secret key belonging to the gateway *GW*. Then, *SA* stores K_{snj} in the sensor node memory SN_j. Thus, each sensor node shared a unique symmetric key K_{snj} with the gateway.

8.4.3.3. *Healthcare professional registration phase*

At first, a user U_i who is a healthcare professional should register at the gateway, through an administration system *SA*, so as to be authorized to access the wireless medical sensor network. The following steps are needed for registration (see Figure 8.3).

– Step 1: U_i chooses an identity ID_i and a password PW_i. Then, they imprint their personal biometric data B_i (fingerprint) on a specific device and calculate $(B^*_i, P_i) = Gen\ (B_i)$. $Gen\ ()$ is a probabilistic generation procedure by the fuzzy extractor. B^*_i represents the biometric key and P_i designates the public reproduction parameter that will be used later to reproduce B^*_i. Then, U_i calculates $BPW_i = h(PW_i\ //\ B^*_i)$ and $IPW_i = h(PW_i\ //\ ID_i)$. The user then sends the message $\{IPW_i, BPW_i\}$ to the system administrator, and this transmits the message received at the gateway *GW*. Indeed, to resist insider attacks as well as password guessing attack (e.g. a brute force attack) and user identity guessing attack, U_i uses a hashing function to ensure that their personal information, such as a password, their personal identity and their biometric data, are masked when they are transmitted to the system administrator and then to the *GW*.

– Step 2: after having received the message from U_i, the gateway *GW* produces a dynamic identity DID_i for U_i and then calculates $A_i = h\ (DID_i\ //\ k\ //\ ID_g)$, $C_i = A_i \oplus BPW_i$, $D_i = DID_i \oplus BPW_i$ and $F_i = A_i \oplus IPW_i$. In fact, the reason to use a dynamic identity DID_i is to strengthen user untraceability. Then, *GW* sends $\{C_i, D_i, F_i, h(\cdot)\}$ to *SA*. The system administrator then sends a smart card containing the information $\{C_i, D_i, F_i, h(\cdot)\}$ and transmits this to U_i via a secure channel (e.g. hand to hand or via a registered post).

– Step 3: after they have received their smart card from the *SA*, the healthcare professional U_i stores P_i in SC_i. So, after this update, the smart card SC_i finally contains parameters $\{C_i, D_i, F_i, P_i, h(.)\}$.

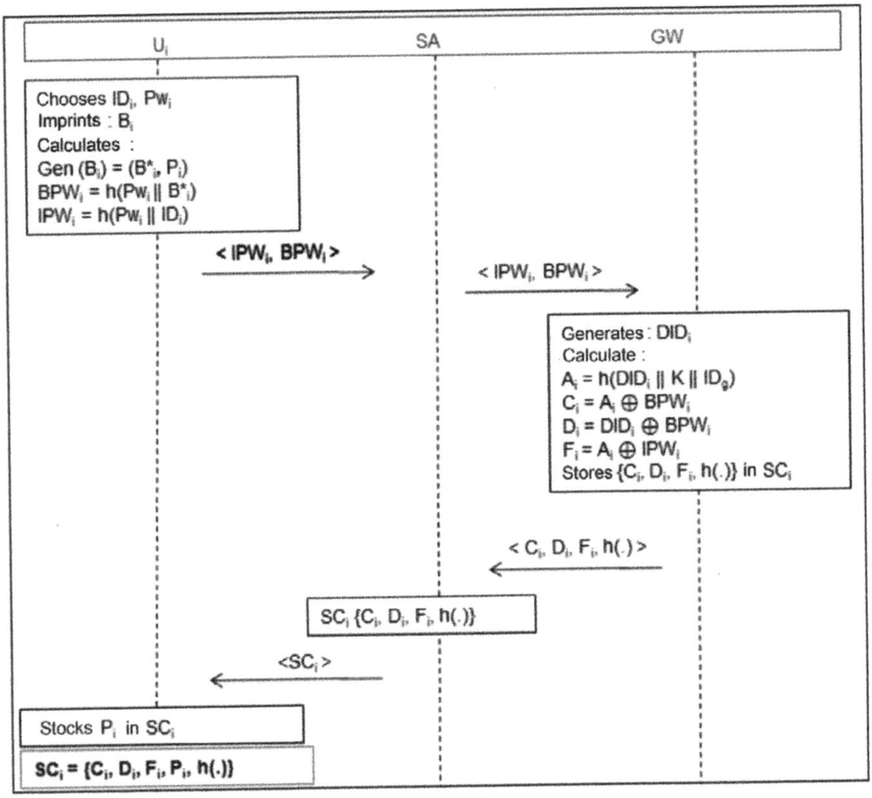

Figure 8.3. *Registration phase of Sammoud* et al. *(2020c). For a color version of this figure, see www.iste.co.uk/chalouf/intelligent.zip*

8.4.3.4. *Login phase*

– Step 4: U_i inserts their own SC_i into the smart card reader and re-enters his ID_i and his PW_i. Then, he imprints his biometric data (fingerprint) B' with the help of a specific sensor (Figure 8.4). Then, the smart card SC_i reproduces $B^*_i = Rep\ (B'_i, P_i)$ where Rep(.) is a deterministic reproduction procedure of the fuzzy extractor. Finally, SC_i calculates $BPW_i = h(PW_i\ //\ B^*_i)$, $A_i = C_i \oplus BPW_i$ and $F'_i = A_i \oplus h(PW_i\ //\ ID_i)$, and compares F'_i with stored F_i. Where they match, U_i passes identity verification and SC_i passes to the following operation stage. Where this does not happen, the connection request is refused.

– Step 5: SC_i recovers $DID_i = D_i \oplus BPW_i$ and generates a random number (nonce) M_i. Then, it calculates $M_1 = M_i \oplus A_i$, $M_2 = h(ID_i\ //\ SID_j)$, $M_3 = E_{h(Mi)}\ (ID_i\ //\ SID_j\ //\ M_2 //\ T_1)$ and transmits $\{DID_i, M_1, M_3\}$ to GW via a public channel (e.g.

the Internet). The random number generated M_i as well as the timestamp T_I helps to protect this approach against replay attacks. Furthermore, M_i is used, masked by a hashing function, as a symmetric key for the message M_3. In fact, this random number is known only by the user. To recover M_i, an attacker needs to be able to guess the user's biometric data and password, which are well protected. This approach therefore guarantees a high level of resistance against user impersonation attack.

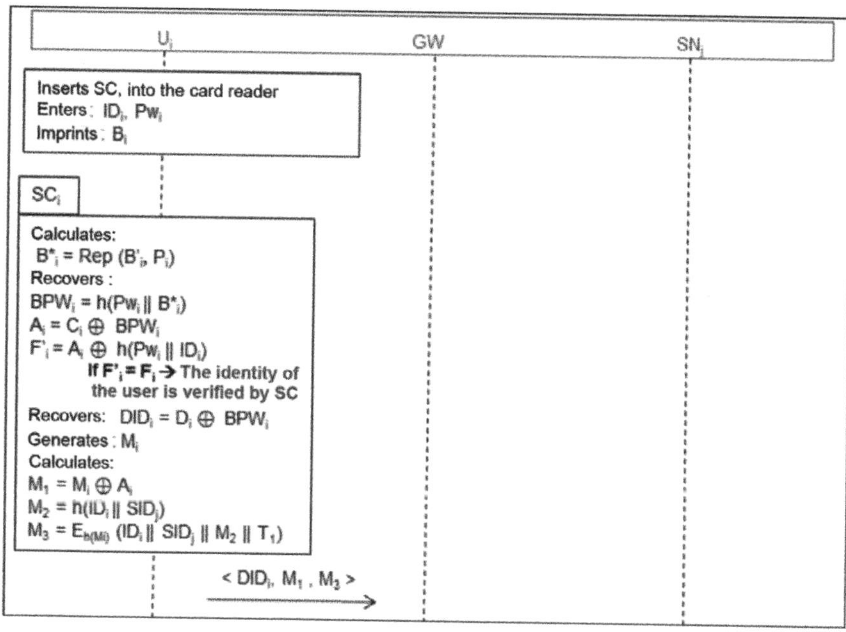

Figure 8.4. *Login phase of Sammoud* et al. *(2020c). For a color version of this figure, see www.iste.co.uk/chalouf/intelligent.zip*

8.4.3.5. *Authentication phase*

To ensure mutual authentication and secure sharing of a session key between U_i and SN_j, the following steps should be carried out (Figures 8.5(a) and 8.5(b)).

– Step 6: after having received the connection request message from U_i, GW should, first of all, recover $M_i = M_1 \oplus h(DID_i \,//\, K \,//\, ID_g)$. Then, GW decrypts M_3 by using $h(M_i)$ as a decryption key to retrieve $\{ID_i \,//\, SID_j \,//\, M_2 \,//\, T_1\}$ and verifies, afterward, if $T_2 - T_1 \leqslant \Delta_T$. Where they do not match, GW ends the session. Otherwise, GW calculates $M'_3 = h(ID_i \,||\, SID_j)$ and compares M'_3 to M_3. Where they do match, U_i is correctly identified by the GW and the following steps will be carried out. Where they do not match, the session is terminated.

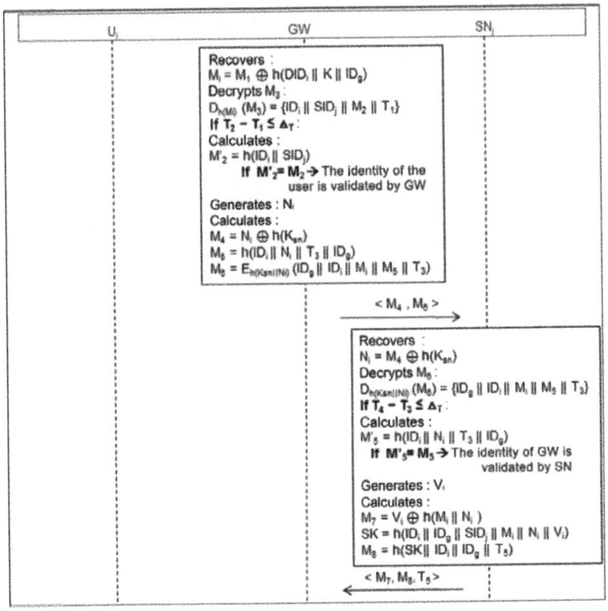

a)

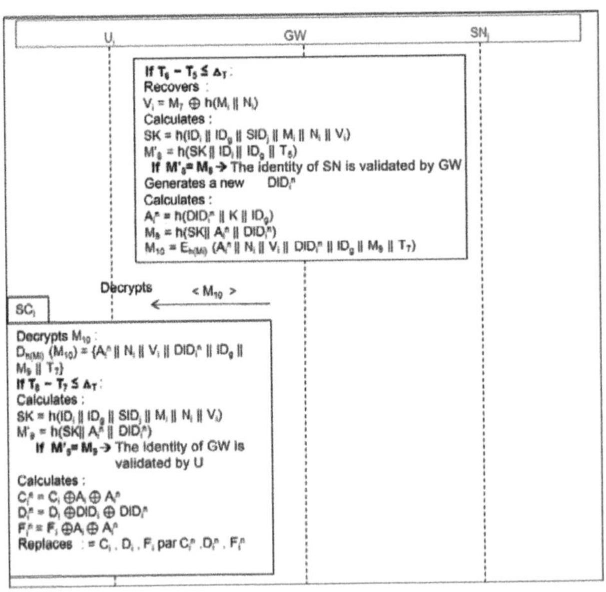

b)

Figure 8.5. *a) Authentication phase of Sammoud* et al. *(2020c); b) Protocol authentication stage of Sammoud* et al. *(2020c) (the continuation). For a color version of this figure, see www.iste.co.uk/chalouf/intelligent.zip*

– Step 7: the gateway *GW* generates a random number (nonce) N_i and calculates $M_4 = N_i \oplus h(K_{sn})$, $M_5 = h\ (ID_i\ //\ N_i\ //\ T_3 //\ ID_g)$ and $M_6 = E_{h(Ksn||Ni)}\ (ID_g\ //\ ID_i\ //\ M_i\ //\ M_5\ //\ T_3)$. Then, *GW* transmits $\{M_4, M_6\}$ to SN_j via a public channel. The random number generated N_i and the timestamp T_3 are used to protect this approach against replay attacks. The message M_6 is encrypted using $h(K_{sn}\ //\ N_i)$ as a symmetric key, where K_{sn} is a symmetric key known only by the gateway *GW* and the sensor SN_j concerned. Furthermore, N_i is very difficult for an attacker to predict. This approach is therefore well protected against gateway impersonation attack.

– Step 8: after having received the message from *GW*, SN_j recovers $N_i = M_4 \oplus h(K_{sn})$ to decrypt M_5 and can extract $\{ID_g\ //\ ID_i\ //\ M_i\ //\ M_5\ //\ T_3\}$. Then, it verifies whether $T_4 - T_3 \leqslant \Delta_T$. Where they do not match, SN_j ends the session. Otherwise, SN_j calculates $M'_5 = h(ID_i\ //\ N_i\ //\ T_3 //\ ID_g)$ and compares M'_5 with the M_5 received. Where they match, *GW* is correctly authenticated by SN_j and other steps will be carried out. Where they do not match, the session will be terminated.

– Step 9: SN_j proceeds to generate the random number (nonce) V_i and calculates $M_7 = V_i \oplus h\ (M_i\ //\ N_i)$, the session key $SK = h(ID_i\ //\ ID_g\ //\ SID_j\ //\ M_i\ //\ N_i\ //\ V_i)$ and $M_8 = h\ (SK\ //\ ID_i\ //\ ID_g\ //\ T_5)$. Then, SN_j transmits $\{M_7, M_8, T_5\}$ to *GW* via a public channel. Use of a random number V_i and the timestamp T_5 helps us to protect this approach against replay attacks. M_7 and M_8 will both be used by *GW* to verify the identity of the sensor SN_j. Both messages will be generated on the basis of data known only by *GW* and SN_j. To do this, we can say that this approach is fairly robust against sensor node impersonation attack.

– Step 10: after having received the message from SN_j, *GW* first verifies whether $T_6 - T_5 \leqslant \Delta_T$. Where they do not match, *GW* ends the session. Otherwise, it recovers $V_i = M_7 \oplus h\ (M_i\ //\ N_i)$ and calculates $SK' = h(ID_i\ //\ ID_g\ //\ SID_j\ //\ M_i\ //\ N_i\ //\ V_i)$ and $M'_8 = h\ (SK\ //\ ID_i\ //\ ID_g\ //\ T_5)$. Then, *GW* compares M'_8 with M_8 received. Where they match, SN_j is correctly authenticated by the *GW* and other stages will be carried out. Where they do not match, the session will be terminated.

– Step 11: *GW* produces a new dynamic identity DID^n_i for the user U_i and calculates $A^n_i = h\ (DID^n_i\ //\ k\ //\ ID_g)$. Then, it calculates $M_9 = h(SK\ //\ A^n_i\ //\ DID^n_i)$ and $M_{10} = E_{h(Mi)}\ (A^n_i\ //\ N_i\ //\ V_i\ //\ DID^n_i\ //\ M_9\ //\ T_7)$ before transmitting M_{10} to U_i via a public channel. The message M_{10} contains the information needed for U_i to be able to generate the symmetric key *SK* and update the parameters stored in their smart card. It is impossible for the attacker to obtain this secret data because the message M_{10} is encrypted with the help of a symmetric key $h(M_i)$ which is the result of hashing the secret information M_i.

– Step 12: after receiving the message M_{10} from GW, U_i (via their smart card) applies the decryption function on M_{10} using $h(M_i)$ as a symmetric key to recover $\{A''_i \,//\, N_i \,//\, V_i \,//\, DID''_i \,//\, ID_g \,//\, M_9 \,//\, T_7\}$. Then, they verify whether $T_8 - T_7 \leqslant \Delta_T$. Where they do not match, U_i ends the session. Otherwise, U_i calculates $SK' = h(ID_i \,//\, ID_g \,//\, SID_j \,//\, M_i \,//\, N_i \,//\, V_i)$ and $M'_9 = h(SK' \,//\, A''_i \,//\, DID''_i)$ and then compares M'_9 to M_9. Where they match, GW is correctly authenticated at U_i and other stages will be carried out. Where they do not match, the session will be terminated. We should note that the secret session key SK generated will later be used to encrypt/decrypt sensitive medical data between the user U_i and the sensor node SN_j. Indeed, this session key SK is unique for each user/gateway/sensor authentication session.

– Step 13: U_i calculates $C''_i = C_i \oplus A_i \oplus A''_i$, $D''_i = DID''_i \oplus BPW_i$ and $F''_i = F_i \oplus A_i \oplus A''_i$ to replace C_i , D_i and F_i stored in the smart card by the new values C''_i , D''_i and F''_i. Indeed, updating the secret parameters stored in the smart card is carried out to guarantee the untraceability of the healthcare professional (user) in this approach.

8.4.3.6. *Password change phase*

This phase is necessary to enable a legal user U_i wanting to update their password in complete security. In this case, the following stages should be executed.

– Step 1: after having inserted his smart card SC_i into a specific card reader, the user U_i enters their identity ID_i, and password PW_i. The user then imprints their biometric data B_i at the corresponding sensor.

– Step 2: to verify the user's identity, just like in the login phase, the smart card SC_i reproduces $B^*_i = Rep\ (B'_i, P_i)$. Then, SC_i calculates $BPW_i = h(PW_i \,//\, B^*_i)$, $IPW_i = h(PW_i \,//\, ID_i)$, $A_i = C_i \oplus BPW_i$ and $F_i = A_i \oplus h(PW_i \,//\, ID_i)$ and compares, afterward, F_i to F_i stored in SC_i. Where they match, the user U_i passes their identity verification and SC_i requires this user to enter the new password. Where they do not match, SC_i refuses the request to change password and ends this stage.

– Step 3: using the new password $PW_{i(new)}$, SC_i should modify the registered parameters C_i, D_i and F_i. Then, SC_i calculates $BPW_{i(new)} = h(PW_{(new)i} \,//\, B^*_i)$, $IPW_i = h(PW_{(new)i} \,//\, ID_i)$, $C_{i(new)} = C_i \oplus BPW_i \oplus BPW_{i(new)}$, $D_{i(new)} = D_i \oplus BPW_i \oplus BPW_{i(new)}$ and $F_{i(new)} = F_i \oplus IPW_i \oplus IPW_{i(new)}$. Finally, SC_i will replace the parameters already stored by the new $\{C_{i(new)}, D_{i(new)}, F_{i(new)}, P_i, h(.)\}$.

8.4.3.7. *Discussion*

The multifactor authentication protocol proposed by Sammoud *et al.* (2020c) is robust in terms of security against known attacks (password attacks, replay attacks, user impersonation attack, gateway impersonation attack and sensor node impersonation attack) while still ensuring low resource consumption. But it still remains vulnerable due to certain drawbacks linked to use of the password and the smart card. Unlike biometrics, a password can be forgotten and the smart card can be lost. The protocol proposed in Sammoud *et al.* (2020c) is robust against password attacks, and even when they know data stored in the smart card, an attacker is still unable to access the user's sensitive data. But it is impossible for the user to be authenticated correctly if they forget their password or do not have their smart card. Biometric characteristics cannot be forgotten nor stolen, nor lost. Because of this, for the remainder of this chapter we present the possibility of using authentication mechanisms based on biometric methods that are processed using machine learning techniques.

8.5. Authentication techniques based on biometrics and machine learning

8.5.1. *Machine learning algorithms*

Several machine learning algorithms are used for identification/authentication. In this section, we present the most powerful machine learning algorithms compared to other algorithms for biometric applications (Yeh *et al.* 2016):

– Neural networks: networks of artificial neurons are inspired by biological neural systems in the human brain (Negnevitsky 2005). They are characterized by their capacity to handle large quantities of information. These networks are formed of a set of neurons working simultaneously. Multilayer (MLP: MultiLayer Perceptron) networks are the most widely used classifiers for neural networks. They are formed using supervised algorithms. RBF (Radial Basis Function) networks derive from the approximation theory of functions that have a powerful architecture (Feedforward). This type of network was introduced for the first time by Hardy (1995), and the corresponding theory was developed by Powell (1999). Afterward, this type of network was known by the "term" neural networks. This type of network uses Gaussian functions. The learning is based essentially on the least squares algorithm and the K-means algorithm.

– Support-vector machines (SVMs): are supervised learning techniques intended to solve problems of regression and discrimination. SVMs were developed during the 1990s and are characterized by their capacity to work with large dimension data.

SVMs use linear separators that maximize the margin between two different classes with the aim of providing an optimal classification (Nguyen and Armitage 2008).

– k-nearest neighbors: the k-nearest neighbors (KNN) method is a supervised learning method that can be used when searching for new biomarkers for diagnostics (Bhatia and Vandana 2010). KNN is a non-parametric method in which the model memorizes observations from all the learning to classify data from all the tests. This machine learning algorithm is called Lazy Learning because it learns nothing during the training stage. To estimate the output linked to a new input, the KNN method consists of considering (identically) the k learning samples whose input is closest to the new input, depending on a distance to be defined (e.g. Euclidean distance) (Bhatia and Vandana 2010).

– Decision tree: the decision tree (DT) method is one of the most popular supervised learning methods for data classification problems. This tool to aid decision-making makes it possible to represent a set of choices in graph form in the shape of a tree. Indeed, a DT models a hierarchy of tests to predict a result. The possible decisions are situated at the extremities (the leaves of the tree) and are reached depending on the decisions at each stage. A DT works by applying very simple logical rules iteratively; these rules are chosen depending on the result of the preceding rule. DTs have the advantages of being simple to interpret and quick to use, of being non-parametric and of requiring very little data pre-processing (Phyu 2009).

8.5.2. *Examples of authentication approaches based on biometrics and machine learning*

These days, mobile phones with touch screens represent an increasingly substantial part of the mobile market. Users often also use their mobile phones to store personal and sensitive data. It is therefore important to protect mobile phones by authenticating legitimate users and by detecting intruders. Authentication on mobile devices often occurs using iris recognition. However, these authentication systems can be attacked. The authors of Gragnaniello *et al.* (2015) used an LBP (Local Binary Pattern) model to evade these attacks. To improve the discrimination capacity of LBP and better explore image statistics, a linear SVM classification schema is proposed in Gragnaniello *et al.* (2015).

In Mondal and Bours (2015), the authors used continuous authentication based on biometrics using mouse dynamics and SVM and ANN algorithms for classification. In Meng *et al.* (2012), the authors proposed a new authentication scheme, based on touch dynamics, that uses a set of behavioral features linked to touch dynamics for a precise authentication. To evaluate their system's performances, they collected and analyzed touch gesture data from 20 Android phone users by comparing several known classifiers. Experimental results show that an optimized neural network classifier (ANN) is well adapted to authenticate different users with an average error rate of around 3% for the selected features.

8.5.3. *Authentication approaches based on ECG and machine learning*

8.5.3.1. *Overall schema for authentication by ECG*

The electrocardiogram (ECG) is a graph representing the electrical activity of the heart. The device used to obtain and display the ECG is called an electrocardiograph (or quite simply an ECG device). This device registers electrical cardiac currents (voltages or potential) by means of selective conductive nodes at the surface of the body and provides a voltage-time graph of the heart's rhythm.

There are two approaches for human identification:

– Fiducial approach: the ECG signal is formed of three waves P, QRS and T (Figure 8.6). The fiducial characteristics describe the peaks and intervals between the three waves. Fiducial methods detect, process and classify these characteristics to identify individuals.

– Non-fiducial approach: to avoid the cost of detecting peaks or intervals, the non-fiducial approach uses signal processing based on the shape of the waves in the frequency domain.

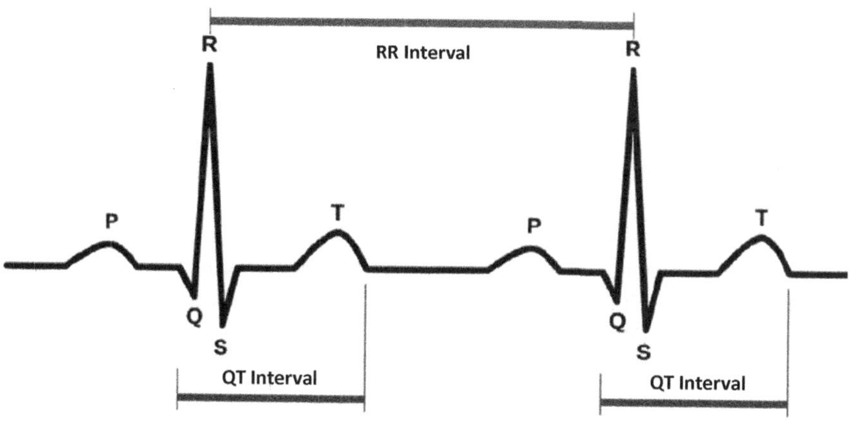

Figure 8.6. *An ECG sequence (Sammoud* et al. *2020a). For a color version of this figure, see www.iste.co.uk/chalouf/intelligent.zip*

In general, authentication by ECG happens in four stages (Figure 8.7).

– Stage 1: acquiring an ECG signal (Figure 8.8). The ECG is acquired via three electrodes placed on the right and left upper limbs and the lower right limb.

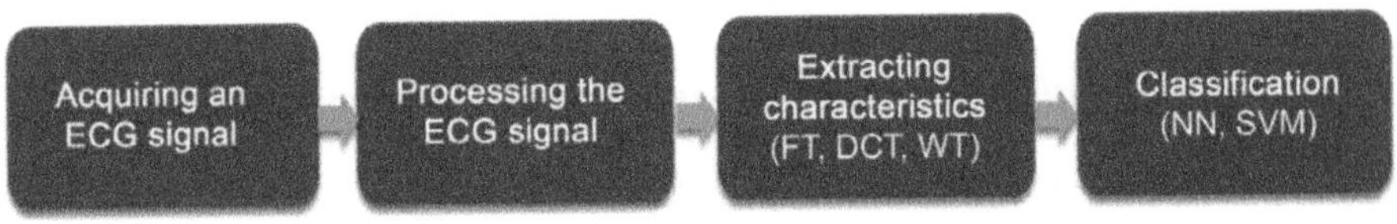

Figure 8.7. *Overall structure of an authentication system based on ECG*

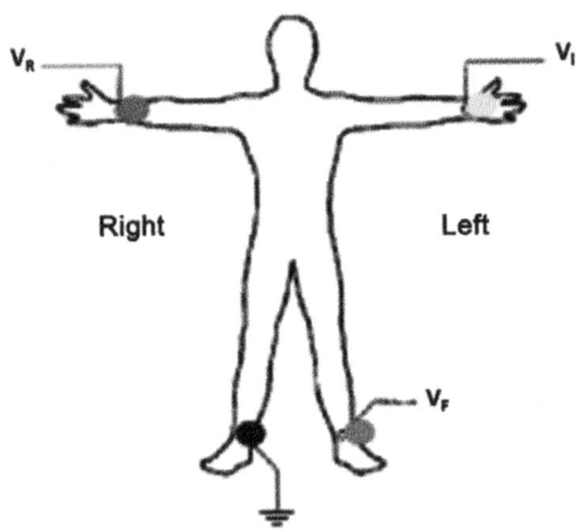

Figure 8.8. *Acquiring an ECG signal. For a color version of this figure, see www.iste.co.uk/chalouf/intelligent.zip*

– Stage 2: pre-processing the ECG signal. Usually, the ECG contains noise. Consequently, a pre-processing stage is carried out using filters (Wang *et al.* 2007).

– Stage 3: extracting characteristics. Several methods have been used such as the Fourier Transform (FT), the Discrete Cosine Transform (DCT) and the Discrete Wavelet Transform (DWT).

– Stage 4: classification. The question is asked: "Is the signal really an ECG graph of an individual or not?" To solve this problem, several approaches try to use machine learning algorithms that can make it possible to predict the data from different types of databases. These ML algorithms are used to discover a set of rules in the database to carry out a precise classification (Wang *et al.* 2007).

8.5.3.2. *Authentication approaches based on ECG*

Use of ECG in biometrics is a recent development. The first method was proposed in 1999 by Biel *et al.* (2001). This method is based on analysis of the ECG using fiducial points and has made it possible to implement the SIMCA (Soft Independent Modeling of Class Analogy) method for classification. Later, in 2006, Wang *et al.* proposed a new method of biometric identification. It is based on fiducial detection using the DCT and the autocorrelation function. This method consists of integrating analytical characteristics and appearance characteristics extracted from the ECG signals for human identification. The latter method, without fiducial detection, called AC/DCT, was proposed by Plataniotis *et al.* (2006).

Recently, researchers have focused on improving classification methods. Several pieces of research are based on machine learning algorithms. Sidek *et al.* (2012) used an MLP algorithm to classify the individuals to be identified. Tantawi *et al.* (2015) also introduced RBF neural networks to classify signals. Tang and Shu (2014) also used Quantum Neural Networks (QNN). Melgani and Bazi (2008) use the SVM classifier and the results obtained confirm that the SVM approach is better than the traditional kNN approach.

Kim *et al.* (2019a) proposed a machine learning framework used for biometric authentication based on the ECG. To evaluate their proposition, they used the three ML techniques: SVM, DT and the neural networks. Al Alkeem *et al.* (2019) proposed a biometric authentication protocol based on ECG and machine learning algorithms. The authors tested the performances of their proposition with two ML protocols: SVM and DT. Kim *et al.* (2019b) proposed a biometric authentication protocol based on machine learning using ECG R-R intervals. The authors analyzed performances of the proposed protocol using two ML techniques: SVM and DT.

Moreover, thanks to the random character of the ECG signal, this sequence can also be used to generate fairly robust symmetric keys, as demonstrated in Sammoud *et al.* (2020a, 2020b). Indeed, ECG represents a good alternative compared to other classical methods based on cryptographic mechanisms to generate cryptographic keys (Sammoud *et al.* 2020a, 2020b).

8.6. Challenges and limits

8.6.1. *Quality of biometric data*

Authentication using biometrics is less precise than that based on what a person knows or possesses. Indeed, comparison of two biometric vectors is expressed in similarity, which never reaches 100%. The quality/precision of sensors has an impact on the quality and performances of biometric authentication systems.

8.6.2. *Non-revocability of biometric data*

Unlike a code, biometric data cannot be modified after falsification or theft. The existence of these vulnerabilities has given rise to a need to protect biometric data.

8.6.3. *Security of biometric systems*

A biometric system can present security failures against attacks on the sensor, attacks on transmission channels, direct attacks on stored data and attacks during the enrollment stage.

In addition, the use of biometrics presents risks in terms of respecting privacy. The capturing, processing and preservation of biometric data can be an invasion of privacy. Biometric data are sensitive in nature and no international standard protects these data. Several solutions can be imagined, such as anonymization or consideration of respect for privacy from the stage of designing the biometric system (Privacy by design).

8.7. Conclusion

Most of the data processed in the IoT environment are sensitive and personal and should be well protected to ensure security and the protection of user privacy.

In this chapter, we have studied the contribution of biometrics and artificial intelligence to strengthening security in the IoT. First, we presented security problems in the IoT. Indeed, our interest was in studying authentication techniques in the IoT, based on biometrics. We have therefore presented multifactor authentication protocols as well as authentication approaches based on biometrics and machine learning. In particular, we studied approaches that use artificial intelligence techniques such as machine learning to ensure authentication based on ECG.

8.8. References

Ahmed, A.A. and Ahmed, W.A. (2019). An effective multifactor authentication mechanism based on combiners of hash function over Internet of Things. *Sensors*, 19(17), 3663.

Al Alkeem, E., Kim, S.K., Yeun, C.Y., Zemerly, M.J., Poon, K.F., Gianini, G., Yoo, P.D. (2019). An enhanced electrocardiogram biometric authentication system using machine learning. *IEEE Access*, 7, 123069–123075.

Ali, R., Pal, A.K., Kumari, S., Sangaiah, A.K., Li, X., Wu, F. (2018). An enhanced three factor based authentication protocol using wireless medical sensor networks for healthcare monitoring. *Journal of Ambient Intelligence and Humanized Computing*, 1–22.

Bhatia, N. (2010). Survey of nearest neighbor techniques. *International Journal of Computer Science and Information Security*, 8(2). arXiv:1007.0085.

Biel, L., Pettersson, O., Philipson, L., Wide, P. (2001). ECG analysis: A new approach in human identification. *IEEE Transactions on Instrumentation and Measurement*, 50(3), 808–812.

Cabra, J.L., Mendez, D., Trujillo, L.C. (2018). Wide machine learning algorithms evaluation applied to ECG authentication and gender recognition. In *2nd International Conference on Biometric Engineering and Applications*. CBEES, Amsterdam.

Dhillon, P.K. and Kalra, S. (2017). A lightweight biometrics based remote user authentication scheme for IoT services. *Journal of Information Security and Applications*, 34, 255–270.

Gragnaniello, D., Sansone, C., Verdoliva, L. (2015). Iris liveness detection for mobile devices based on local descriptors. *Pattern Recognition Letters*, 57, 81–87.

Hamidi, H. (2019). An approach to develop the smart health using Internet of Things and authentication based on biometric technology. *Future Generation Computer Systems*, 91, 434–449.

Hardy Sr., R.L. (1995). U.S. Patent No. 5,452,863. Patent, Washington.

Kim, S.K., Yeun, C.Y., Damiani, E., Lo, N.W. (2019a). A machine learning framework for biometric authentication using electrocardiogram. *IEEE Access*, 7, 94858–94868.

Kim, S.K., Yeun, C.Y., Yoo, P.D. (2019b). An enhanced machine learning-based biometric authentication system using RR-Interval Framed Electrocardiograms. *IEEE Access*, 7, 168669–168674.

Lee, H., Kang, D., Ryu, J., Won, D., Kim, H., Lee, Y. (2020). A three-factor anonymous user authentication scheme for Internet of Things environments. *Journal of Information Security and Applications*, 52, 102494.

Li, X., Peng, J., Obaidat, M.S., Wu, F., Khan, M.K., Chen, C. (2019). A secure three-factor user authentication protocol with forward secrecy for wireless medical sensor network systems. *IEEE Systems Journal*, 14(1), 39–50.

Melgani, F. and Bazi, Y. (2008). Classification of electrocardiogram signals with support vector machines and particle swarm optimization. *IEEE Transactions on Information Technology in Biomedicine*, 12(5), 667–677.

Meng, Y., Wong, D.S., Schlegel, R. (2012). Touch gestures based biometric authentication scheme for touchscreen mobile phones. In *International Conference on Information Security and Cryptology*. ICISC, Seoul.

Mondal, S. and Bours, P. (2015). A computational approach to the continuous authentication biometric system. *Information Sciences*, 304, 28–53.

Negnevitsky, M. (2005). *Artificial Intelligence: A Guide to Intelligent Systems*, 2nd edition. Pearson Education, London.

Nguyen, T.T. and Armitage, G. (2008). A survey of techniques for internet traffic classification using machine learning. *IEEE Communications Surveys & Tutorials*, 10(4), 56–76.

Ometov, A. and Bezzateev, S. (2017). Multi-factor authentication: A survey and challenges in V2X applications. In *2017 9th International Congress on Ultra Modern Telecommunications and Control Systems and Workshops (ICUMT)*. IEEE, Munich.

Ometov, A., Bezzateev, S., Mäkitalo, N., Andreev, S., Mikkonen, T., Koucheryavy, Y. (2018). Multi-factor authentication: A survey. *Cryptography*, 2(1), 1.

Phyu, T.N. (2009). Survey of classification techniques in data mining. *Proceedings of the International MultiConference of Engineers and Computer Scientists*, 1(5), 727–731.

Plataniotis, K.N., Hatzinakos, D., Lee, J.K. (2006). ECG biometric recognition without fiducial detection. In *2006 Biometrics Symposium: Special Session on Research at the Biometric Consortium Conference*. IEEE, Baltimore.

Powell, M.J. (1999). Recent research at Cambridge on radial basis functions. In *New Developments in Approximation Theory*, Müller, M.W., Buhmann, M.D., Mache, D.H., Felten, M. (eds). Birkhäuser, Bâle.

Sammoud, A., Chalouf, M.A., Hamdi, O., Montavont, N., Bouallegue, A. (2020a). A new biometrics-based key establishment protocol in WBAN: Energy efficiency and security robustness analysis. *Computers & Security*, 96, 101838.

Sammoud, A., Chalouf, M.A., Hamdi, O., Montavont, N., Bouallegue, A. (2020b). A secure three-factor authentication and biometrics-based key agreement scheme for TMIS with user anonymity. In *2020 International Wireless Communications and Mobile Computing (IWCMC)*. IEEE, Limassol.

Sammoud, A., Chalouf, M.A., Hamdi, O., Montavont, N., Bouallegue, A. (2020c). A secure and lightweight three-factor authentication and key generation scheme for direct communication between healthcare professionals and patient's WMSN. In *2020 IEEE Symposium on Computers and Communications (ISCC)*. IEEE, Rennes.

Sidek, K.A., Khalil, I., Smolen, M. (2012). ECG biometric recognition in different physiological conditions using robust normalized QRS complexes. In *2012 Computing in Cardiology*. IEEE, Krakow.

Tang, X. and Shu, L. (2014). Classification of electrocardiogram signals with RS and quantum neural networks. *International Journal of Multimedia and Ubiquitous Engineering*, 9(2), 363–372.

Tantawi, M.M., Revett, K., Salem, A.B., Tolba, M.F. (2015). A wavelet feature extraction method for electrocardiogram (ECG)-based biometric recognition. *Signal, Image and Video Processing*, 9(6), 1271–1280.

Wang, Y., Plataniotis, K.N., Hatzinakos, D. (2006). Integrating analytic and appearance attributes for human identification from ECG signals. In *2006 Biometrics Symposium: Special Session on Research at the Biometric Consortium Conference*. IEEE, Baltimore.

Wang, Y., Agrafioti, F., Hatzinakos, D., Plataniotis, K.N. (2007). Analysis of human electrocardiogram for biometric recognition. *EURASIP Journal on Advances in Signal Processing*, 2008(1), 148658.

Xiao, L., Wan, X., Lu, X., Zhang, Y., Wu, D. (2018). IoT security techniques based on machine learning: How do IoT devices use AI to enhance security? *IEEE Signal Processing Magazine*, 35(5), 41–49.

Yeh, K.H., Su, C., Hsu, C.L., Chiu, W., Hsueh, Y.F. (2016). Transparent authentication scheme with adaptive biometrie features for IoT networks. In *2016 IEEE 5th Global Conference on Consumer Electronics*. IEEE, Kyoto.

9

Dynamic Identity and Access Management in the IoT: Blockchain-based Approach

Léo MENDIBOURE[1], Mohamed-Aymen CHALOUF[2] and Francine KRIEF[3]

[1] LaBRI, University of Bordeaux, Talence, France

[2] IRISA, University of Rennes 1, Lannion, France

[3] LaBRI, Bordeaux INP, Talence, France

9.1. Introduction

The Internet of Things (IoT) is developing today in many sectors such as energy, telecommunications, e-health or transport systems (Chen *et al.* 2014). This advent of the IoT has led to a significant increase of the number of connected objects: watches, cameras, lighting systems, electric meters, etc.

To guarantee the proper functioning of applications in the IoT, it is vital to make these connected objects secure (Xiao *et al.* 2018). Indeed, if these objects are compromised, it could lead to many types of attack: denial of service, ransomware, industrial espionage, etc. In this security process, identity and access management is an essential first step. By authenticating communicating objects and by controlling access to resources (storage, calculation and communication links), it could be possible to limit the risk of objects being compromised (Fang *et al.* 2020a).

However, classical authentication and access control solutions could be complex to apply in the IoT environment (Xiao *et al.* 2018). Indeed, many connected objects only have limited capabilities and it would not be possible to deploy cryptographic

Intelligent Security Management and Control in the IoT,
coordinated by Mohamed-Aymen CHALOUF. © ISTE Ltd 2022.

solutions that are costly in terms of bandwidth, energy and calculating capacity (Zhou *et al.* 2017). Similarly, the mobility of these objects, their significant number and the need for short authentication delays represent obstacles to using techniques based on key exchange (Fang *et al.* 2020a).

This is why new solutions have to be developed for the authentication and access control of connected objects. Among the methods envisaged, the use of an intelligent approach, based on artificial intelligence (AI) techniques, has appeared as a promising solution (Xiao *et al.* 2018). Indeed, AI techniques could make it possible to guarantee short authentication delays and access control as well as efficient management of mobile and heterogeneous IoT devices. In addition, some work has demonstrated that lightweight AI approaches, adapted to the resources available at the level of connected objects, could be designed (Fang *et al.* 2020b).

Nevertheless, many directions for improvement still need to be considered to deploy such intelligent approaches for identity and access control management: trust establishment, load balancing, anonymization, etc. To deal with these different problems, it may be pertinent to connect intelligent identity and access management with an emerging technology: blockchain (Zhu and Badr 2018). Indeed, blockchain provides a decentralized and secure environment for storing and exchanging data. It could therefore be used to meet the needs of existing solutions using AI for access and authentication control (Singh *et al.* 2020).

Thus, in this chapter, we propose demonstrating how intelligent identity and access management in the IoT could be improved thanks to use of blockchain technology.

To do this, we begin by presenting, in section 9.2, the operation of some intelligent approaches for authentication and access control in the IoT. We also introduce the principle of blockchain technology (section 9.2.2). Second, in section 9.3, we describe an IoT architecture combining blockchain and AI and highlight the benefits that this architecture could present for intelligent identity and access management. Finally, in section 9.4, we identify research perspectives for intelligent identity and access management in the IoT based on blockchain.

9.2. Context

In this section, we present two elements essential for the understanding of this chapter: first, the benefit of intelligent identity and access management in the IoT (section 9.2.1); second, the benefits that blockchain could bring to intelligent identity and access management (section 9.2.2).

9.2.1. *Intelligent identity and access management*

The characteristics of connected objects (heterogeneity, mobility, low calculation/storage/communication capacities, substantial number) complicate the deployment of traditional security solutions, particularly those based on the use of key public infrastructures.

There are three main stumbling points, identified by the authors of Fang *et al.* (2020a). First of all, the latency and high calculation cost induced by traditional authentication and access control mechanisms, which seem incompatible with the characteristics of connected objects (energy, capacities, etc.). Then, the lack of dynamism in these approaches, which makes it impossible to secure communications between objects continuously. Finally, the difficulty of managing keys (generation, distribution, removal). Indeed, in an open environment such as the IoT, this management could lead to many security and privacy protection problems.

Also, the definition of new lightweight, quick and dynamic solutions has emerged as essential. For these, the use of intelligent approaches, based on AI techniques, is a pertinent solution. Indeed, a dynamic identity management could make it possible to approach three limitations identified in Fang *et al.* (2020a). First, intelligent identity management could be achieved closest to the objects, that is, at gateways and routers. Thus, this approach could limit the calculation and communication overhead at the device level. Similarly, this approach could reduce authentication and access control delays (proximity, use of communication histories, etc.). In addition, real-time learning, permitted by some AI techniques, could guarantee significant adaptability. Thus, continuous security could be assured for the objects. Finally, the use of AI techniques could guarantee security without needing to transmit sensitive information such as keys. The risks of communications being intercepted could be limited and security strengthened.

The authors of Xiao *et al.* (2018) propose a classification of existing work using an intelligent approach for authentication and access control for objects in IoT environments. This shows that many approaches (supervised learning, unsupervised learning and deep learning), applied to different layers (physical and link) and using different inputs (signal to noise ratio, MAC addresses and packages received), have already been considered. These approaches, like the solutions described by the authors of Chatterjee *et al.* (2018), Das *et al.* (2018) and Punithavathi *et al.* (2019), meet different issues identified. They represent a significant evolution in the way we think about management of identities and access control in the IoT and could enable the development of lightweight, fast and flexible solutions.

Nevertheless, as noted by the authors of Singh *et al.* (2020), different improvements should still be considered to offer efficient solutions for the intelligent identity and access control management in the IoT. Among the most significant improvements required, discussed in the rest of this chapter, we can highlight the trust level granted to the processed data, the protection of user privacy or the calculation distribution needed to deploy intelligent solutions.

9.2.2. Blockchain

Blockchain is now the most popular distributed ledger approach (Zhu and Badr 2018). A distributed ledger, like blockchain, enables different nodes to exchange direct and secure data without needing the intervention of a trusted third party. This is the main objective of distributed ledger technologies: establishing trust between different nodes without requiring the use of an expensive central control body and requiring a substantial level of security (a single point of attack).

Operation of blockchain is based on different elements (see Figure 9.1). First of all is a chain of blocks, which corresponds to the data structure used for storing data in a blockchain network. This chain of blocks is present at each of the nodes and contains all the data exchanged in blockchain network. Then, a pair-to-pair (P2P) network that enables direct communication between blockchain nodes. A third element is the consensus mechanism. This is a decentralized protocol making it possible to ensure the security of the network and especially for each node to check the reliability of the information exchanged in blockchain network. Finally, a last integrated element of blockchain regroups the decentralized applications. These correspond to IT programs called smart contracts, deployed within blockchain. These smart contracts simplify access to data in a blockchain environment: visualization, modification, etc. The combination of these four elements (blockchains, P2P network, consensus mechanism and decentralized applications) guarantees proper operation of blockchain technology.

The privacy protection, the load distribution and the trust establishment are some of the benefits offered inherently in the blockchain approach. Indeed, without a central control body and in a decentralized architecture, it is necessary for each node to participate in the proper functioning of the network and in verifying the information exchanged (calculation decentralization). We can also note that this decentralized approach guarantees significant scaling up (Mendiboure *et al.* 2020a). This is an important benefit in the IoT environment. In addition, with blockchain, users are anonymized to protect their privacy. Finally, blockchain, thanks to decentralized applications, could allow efficient management of the data stored within blockchain while still guaranteeing its security (immutability, validity) and so reinforcing confidence.

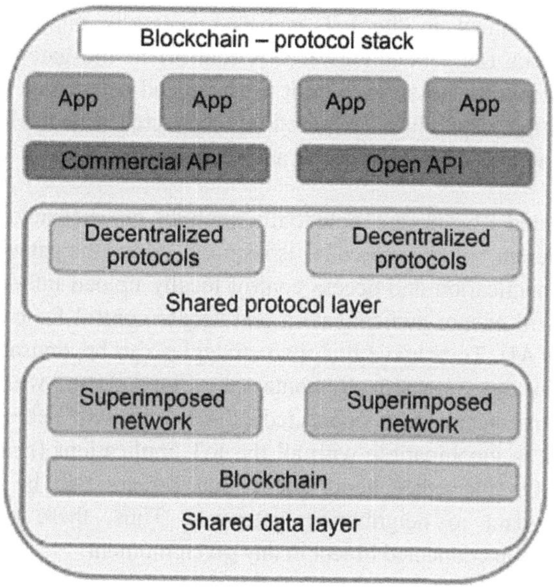

Figure 9.1. *Blockchain architecture (Che et al. 2019). For a color version of this figure, see www.iste.co.uk/chalouf/intelligent.zip*

Consequently, blockchain, used efficiently, could offer a response to the limits identified for intelligent identity and access management in the IoT: confidence, privacy, load distribution, etc. In the remainder of this chapter, we will study the solutions proposed up to now considering use of blockchain for intelligent identity and access management in the IoT.

9.3. Blockchain for intelligent identity and access management

In this section, we highlight the benefits that blockchain could bring to intelligent identity and access management in the IoT. To do this, we show in section 9.3.1 how blockchain could be integrated into the architecture of intelligent identity and access management. Then, we present in section 9.3.2 different works considering the use of blockchain to meet the current limits of intelligent identity and access management in the IoT: privacy protection, intelligent load distribution, trust management, the development of new intelligent solutions and the explainability of the AI process.

9.3.1. *A new architecture integrating blockchain*

As shown in Figure 9.2(a) and as indicated in section 9.2.1, intelligent identity and access management in the IoT is based generally on the deployment of intelligent functions (authentication, access control) at the gateways and routers (edge AI). This makes it possible to limit the overhead calculation cost and storage and communication capabilities required at connected objects level while still deploying the functions of authentication and access control closest to users.

Since connected objects are potentially mobile, an exchange of information between the different "intelligent cells" is required. Thus, the gateways responsible for ensuring authentication and access control locally upload information to Cloud servers, which will ensure authentication and access control for connected objects globally (central AI). To achieve this, two strategies can be envisaged. On the one hand, the gateway can systematically contact the Cloud servers when it is unable to authenticate/control access to a connected object. This will necessarily lead to a latency time that is incompatible with all the IoT applications (health, self-driving vehicles, etc.). On the other hand, the Cloud servers can be responsible for distributing the data to neighboring gateways. Thus, these will be able to authenticate a mobile connected object in any given moment.

As noted in section 9.2.1, this approach presents different limits. First of all, a centralized data processing is needed at the Cloud level, which necessarily leads to a latency time and uploading of a substantial volume of data (central AI). Then, nothing guarantees that an "intelligent cell" will be a trusted cell and consequently that the data it uploads will be reliable and usable by neighboring cells for the authentication and access control of connected objects. Finally, protection of user privacy is not guaranteed.

This is why a new architecture integrating blockchain has been proposed in a number of papers (Qian *et al.* 2018; Rathore *et al.* 2019b; Singh *et al.* 2020). The current architecture, presented in Figure 9.2(b), is based on deployment of blockchain nodes at the level of the gateways and routers. The primary role of these blockchain nodes is to harvest and store all the information available to the IoT gateway that is useful to identity and access management. This makes it possible, first of all, to ensure that these data will still be available for intelligent identity and access management processes. This then makes possible a direct and secure transmission of data between the different "intelligent cells", thanks to blockchain network. This approach therefore makes it possible to eliminate the central authority (central AI) that can cause network congestion and latency time. Each local "intelligent cell" therefore has at any moment enough information to authenticate/control the access of IoT objects without requiring the intervention of an external entity.

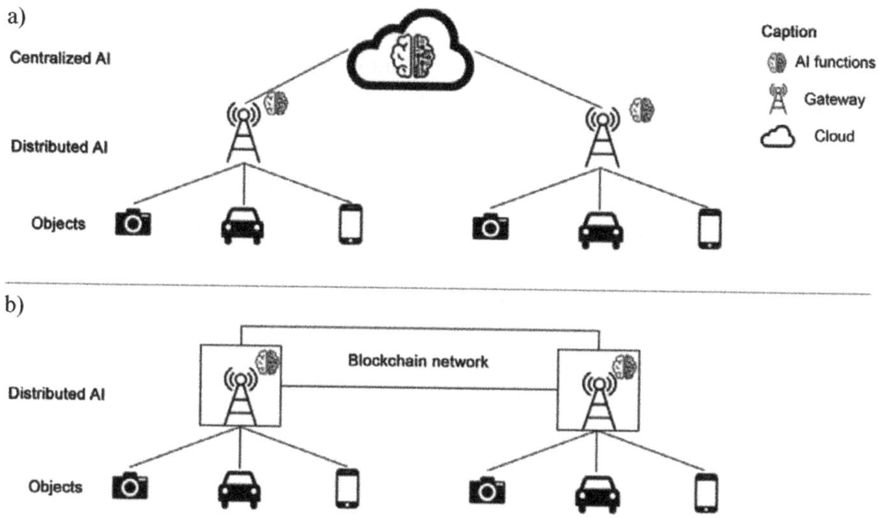

Figure 9.2. *a) Traditional intelligent identity and access management; b) Blockchain-based intelligent identity and access management*

9.3.2. *The different benefits*

Integration of blockchain into the architecture of intelligent identity and access management could lead to a significant evolution in the way we think about and deploy the intelligent services required to authenticate and control the access of objects in an IoT environment (see section 9.3.1). In this section, we show how this evolution could be used to meet the current limitations of intelligent identity and access management (see section 9.2.1): calculation centralization, lack of trust and lack of users' privacy protection. We also highlight the way in which blockchain could enable the advent of new and understandable AI techniques.

9.3.2.1. *Load distribution*

As noted by the authors of Outchakoucht *et al.* (2017), the use of a central entity for intelligent entity and access management leads to different problems, especially latency, the presence of a single point of attack and the difficulty of deploying end-to-end security solutions in the IoT environment. Consequently, the development of solutions allowing distribution of the whole calculation process between the different gateways has emerged as a necessity.

This is why many works, like Outchakoucht *et al.* (2017), Chen *et al.* (2018), Liu *et al.* (2018), Zhou *et al.* (2019) and Cui *et al.* (2020), have proposed solutions combining blockchain and AI techniques for the intelligent identity and access management in the IoT. Blockchain, based on a P2P network, guarantees a direct exchange of data between the different local "intelligent cells". It therefore permits load distribution between the different entities.

Nevertheless, to apply this solution, two essential points should be considered. First, data security should be ensured from end-to-end. Otherwise, proper operation of the authentication and access control process could not be guaranteed. Then, the exchange of data between the different nodes of blockchain should be managed efficiently. Otherwise, the blockchain network could be overloaded. For this reason, existing work focusing on the load distribution for intelligent identity and access services considers these two questions.

The idea of securing data from end-to-end has been tackled especially by Outchakoucht *et al.* (2017), Chen *et al.* (2018), Liu *et al.* (2018) and Zhou *et al.* (2019). These different works aim to guarantee that data added to blockchain (see section 9.2.2) really are data added by the gateway and that no malicious entity has sought to modify it. To do this, different mechanisms based on cryptographic solutions (keys, hashing) are described and their performances evaluated (Chen *et al.* 2018). Beyond these solutions, Chen *et al.* (2018) and Liu *et al.* (2018) also highlight the way in which operation of blockchain technology guarantees the validity of information exchanged between the different nodes of blockchain. Consequently, application of these solutions could guarantee the validity of the data exchanged and, consequently, the efficient functioning of authentication and access control processes.

The authors of Cui *et al.* (2020) are interested in an optimal identity and access management. To do so, they consider a blockchain network with several levels: local networks (corresponding to a gateway or set of gateways) and a global network (interconnecting all the gateways). A fixed object, located in a defined area, will not need to be authenticated globally. Consequently, it will only be possible to store information from this object locally. Thus, this approach limits the amount of data stored in the global blockchain network. To guarantee the proper functioning of this solution, the authors describe the storage and authentication mechanisms proposed in this paper. Application of this solution could permit more efficient use of resources (energy, storage and communication) and greater scaling up.

9.3.2.2. *Establishing trust*

An efficient and secure distribution of data (see section 9.3.2.1) is not enough to guarantee the proper functioning of intelligent authentication and access control processes. Indeed, a malicious or infected gateway could transmit erroneous data and make it impossible to authenticate and control access to mobile connected objects. Consequently, establishing trust between different entities, in this case the gateways, is essential.

One solution currently being considered to establish trust between different entities consists of systematically analyzing the behavior of each of these entities and detecting abnormal behaviors. To deploy such solutions, blockchain represents a pertinent approach. Indeed, on the one hand it permits storage of all the data, generated by the entities (IoT gateways), useful for authentication and control of access to objects. On the other hand, thanks to smart contracts (see section 9.2.2), it could make it possible to analyze this data systematically, securely and autonomously. Consequently, blockchain could be used to ensure trust in distributed authentication and access control processes.

Other research works, such as Kang *et al.* (2018), Mendiboure *et al.* (2018) and Awais Hassan *et al.* (2019), sought to couple blockchain and AI techniques for managing identities and access. These papers are based on similar approaches. For each entity (gateway), a trust index, whose value is stored in blockchain, is defined. An algorithm (subjective logic, fuzzy logic), deployed as a smart contract within this same blockchain, is then used to analyze all the data generated by each of the entities. Depending on the results of this algorithm, the trust index linked to each entity is then increased or lowered. The higher the trust index, the more reliable the data generated by this entity is considered to be. The authors of Mendiboure *et al.* (2018) go further and propose limiting the rights of entities when their trust index is too low. This could be used to limit the quantity of erroneous data stored in blockchain ledger.

Thus, the use of blockchain technology for the intelligent identity and access management makes it possible to couple an effective and secure exchange of information (see section 9.3.2.1) and an effective trust management. The level of trust granted to a given entity, here a gateway, can be considered by each "intelligent cell" in the authentication and access control process for a mobile object arriving in this new cell. This therefore guarantees greater reliability for the authentication and access control process.

9.3.2.3. *Anonymizing users*

Beyond the problems associated with trust establishment and with load balancing, the implementation of solutions based on AI techniques also poses problems with the protection of user privacy (Jameel *et al.* 2020). Indeed, the development of such solutions requires the storage and exchange of data that could be used by malicious entities, for example to trace users. Therefore, the protection of user privacy and the anonymization of this data form a major challenge.

Blockchain thanks to data protection could guarantee a high level of protection against attacks by external malicious entities. Indeed, these will not be able to access encrypted data. Nevertheless, blockchain technology does not inherently enable the system to protect itself against an entity belonging to blockchain network. Indeed, as shown by the authors of Khalilov and Levi (2018) and Feng *et al.* (2019), the pseudo-anonymization permitted by blockchain is not enough to protect user privacy. The untraceability of data stored in blockchain ledger should also be assured.

This is why a great deal of the work combining AI and blockchain has defined solutions aiming to guarantee the anonymity and untraceability of data stored in blockchain ledger (Shen *et al.* 2019; Jameel *et al.* 2020; Kuo *et al.* 2018; Zhao *et al.* 2020). Different strategies are employed to achieve this protection of privacy. In Zhao *et al.* (2020), the authors consider the use of differential privacy techniques (Abadi *et al.* 2016). These, thanks to the introduction of chance, also called noise, make it possible to limit the authentication risks of the entities with which these data originate without however harming statistical use of these data. Thus, the proper functioning of AI techniques is assured while still guaranteeing privacy protection. In Shen *et al.* (2019), a cryptosystem by Pallier (Jost *et al.* 2015) is used. This enables AI techniques to be applied on encrypted data without access to these data being made possible. Finally, a last solution, considered in Jameel *et al.* (2020) and Kuo *et al.* (2018), consists of adding anonymity to the properties of blockchain ledger. Thus, by offering data pseudo-anonymity and untraceability, blockchain ledger could make it possible to protect the privacy of network users.

Consequently, blockchain technology used alone (Jameel *et al.* 2020; Kuo *et al.* 2018) or associated with other mechanisms (Shen *et al.* 2019; Zhao *et al.* 2020) seems to be able to ensure data storage and privacy protection (anonymity, untraceability) in the IoT environment.

9.3.2.4. *The advent of new intelligent approaches*

Intelligent identity and access management in the IoT, and more broadly the deployment of solutions based on AI techniques, is now undergoing significant advances.

The development of collaborative (Yang *et al.* 2019b), or distributed learning, is one of the main evolutions considered. Collaborative learning consists of training a single algorithm in a decentralized manner at the level of each entity, without these entities needing to exchange data. Once this algorithm is trained, the different entities pool the model generated to propose an optimal solution. This approach therefore makes it possible to limit the quantity of data exchanged between entities and so to reduce the latency and the excess cost of communication and storage while still protecting user privacy.

Nevertheless, the deployment of solutions based on a collaborative approach presents different limits, especially the securing of exchanges and the establishment of trust (Bao *et al.* 2019). Indeed, as shown in the previous sections, the deployment of intelligent solutions based on blockchain guarantees high levels of security and trust. This is why many papers have proposed functional solutions to deploy a collaborative learning in blockchain environment (Bao *et al.* 2019; Rathore *et al.* 2019a). These solutions aim, especially, to permit a secure exchange of models and selection of the optimal model depending on the level of trust granted to the objects. Thus, blockchain could permit the advent of collaborative learning, and more broadly of distributed learning (Salah *et al.* 2019). This could guarantee an intelligent identity and access management as well as strengthened privacy protection at low additional cost (in terms of network, storage, latency, etc.).

Beyond collaborative learning, as noted by the authors of Salah *et al.* (2019), blockchain could also be applied to other emerging learning techniques such as the definition of hybrid learning models (Fang *et al.* 2020b) or learning with Augmented Data (AD) (Laskin *et al.* 2020). Indeed, in these contexts, the establishment of trust and the immutability of data are essential elements (Salah *et al.* 2019).

Thus, in addition to meeting the current limits of mechanisms for intelligent identity and access management (see section 4.2.1), blockchain could also improve performances of the intelligent authentication and access control process.

9.3.2.5. *Explainability of the decision process*

One current limit of AI algorithms is the impossibility of explaining these algorithms' decisions. This lack of explainability is especially true for deep learning algorithms. Indeed, even when these algorithms' inputs and outputs are known, it is impossible to explain their decision-making process.

Consequently, despite these algorithms' high performances, it now seems complex to deploy them in critical environments (Nassar *et al.* 2020) such as connected healthcare or self-driving cars. Indeed, without being able to make sure that the decision-making is correct and that no erroneous data have been integrated into the decision-making process, it seems impossible to use these AI algorithms for making critical decisions.

This is why a great deal of research focuses on the explainability of the decision-making process of AI algorithms. We therefore speak of XAI (*eXplainable Artificial Intelligence*) (Gunning 2017). The aim of XAI is to be able to explain the operation of AI algorithms, considered up until now as "black boxes", or, by default, to be able to guarantee that these algorithms' decision-making is correct.

To ensure that a decision-making process is correct, the XAI approach consists generally of making sure that biases and erroneous information that may influence the decision-making process are few, or indeed non-existent (Nassar *et al.* 2020). Indeed, in the absence of bias, the correct functioning of AI algorithms could be guaranteed.

To achieve this, the deployment of XAI solutions in a blockchain environment is pertinent and has been proposed in different papers (Dinh and Thai 2018; Calvaresi *et al.* 2019; Nassar *et al.* 2020). In this context, the advantages of blockchain (data security, traceability of sources, immutability and establishment of trust) are put to the service of the XAI. Thus, thanks to the environment of trust provided by blockchain, it becomes possible to make sure that the inputs provided to AI algorithms are valid and reliable data.

By considering that the operation of the AI algorithms employed has been verified in advance, the combination of blockchain and the XAI ensures that the decision-making processes of AI algorithms will be correct. Thus, blockchain technology could enable the adoption of AI algorithms by reinforcing their level of trust and resilience.

9.4. Challenges

As shown in section 9.3, the use of blockchain technology for intelligent identity and access management in the IoT could bring many benefits: load distribution, trust establishment, privacy protection, development of new AI approaches and AI algorithm explainability.

Nevertheless, to enable the deployment of blockchain-based solutions for intelligent identity and access management in a real environment, many barriers remain to be lifted, especially the scaling up of blockchain, the security of blockchain, the energy consumption and the definition of consensus algorithms designed for the AI environment.

9.4.1. *Scaling up*

Scaling up is an important first axis for improvement of blockchain-based mechanisms for intelligent identity and access management. As has been shown by the authors of Nasir *et al.* (2018), it is important to limit the size of blockchain network to guarantee high performances (latency and bandwidth). In addition, this limitation of block size is essential for allowing its deployment at the level of IoT platforms with limited capacities (in storage, calculation and communication).

This is why, like the solutions proposed in Cui *et al.* (2020) and Mendiboure *et al.* (2020b), a great deal of research focuses today on breaking down blockchain network into a set of sub-networks: on the one hand, a global network and, on the other hand, local networks. This approach, used effectively, could make it possible to store only a limited quantity of information in the global network shared between all the nodes and thus to reduce its size. Nevertheless, the solutions proposed up until now have only been experimental. It is necessary to implement them, deploy them and evaluate them in a real environment.

One solution consists of establishing direct communications between blockchain clients and of storing only a limited quantity of data in blockchain network. This approach, called a state channel, is promising and could permit a significant improvement in performances (McCorry *et al.* 2019). Nevertheless, the application of state channels in an environment such as the IoT would involve developing this technology, especially to manage the mobility and discontinuity of communications (Mendiboure *et al.* 2020b).

9.4.2. *Blockchain security*

Identity and access management can only be possible in an environment whose security is guaranteed. Indeed, any security failure could lead to this identity and access management system being inaccessible.

This is why it is essential to reinforce the security of blockchain technology to ensure that it is not vulnerable to attacks. To do this, three major open axes should be considered (Salah *et al.* 2019). First, it is necessary to focus on the security of the consensus algorithms considered. Indeed, the existing solutions are vulnerable to so-called 51% attacks (Gervais *et al.* 2016). The development of new mechanisms making it possible to strengthen this security is necessary. Another important axis of progress is the securing of smart contracts. Any vulnerability in these smart contracts could result in inefficiency in blockchain network. Thus, a great deal of research, like Luu *et al.* (2016), focuses on detecting vulnerabilities in smart contracts. Nevertheless, the development of new dynamic solutions that propose counter measures for the vulnerabilities identified is needed. A final important point is the definition of secure oracles for blockchain technology. These oracles, making it possible to interact with the outside world, should especially make the link between blockchain and AI techniques. It is therefore essential to secure them. To achieve this, solutions based on inter-oracle validation approaches and the establishment of trust are now being studied (Salah *et al.* 2019).

9.4.3. *Energy consumption*

In our current society, the optimization of energy consumption associated with a given process is now a major challenge. Indeed, the consensus algorithm used today by most blockchain technologies, PoW (Proof of Work), leads to a substantial and unnecessary additional calculation cost (Gervais *et al.* 2016).

This is why much research focuses on defining new consensus algorithms with low energy cost. Among the solutions highlighted in blockchain projects, we find PoS (Proof of Stake), DPoS (Delegated PoS) or PoA (Proof of Authority). Nevertheless, the security level of these different mechanisms is being questioned (Lin and Liao 2017). In addition, some of these solutions, especially PoS and DPoS, are intrinsically linked to the existence of a crypto-currency system. Therefore, they do not seem to be applicable to the identity and access management. Consequently, the development of new, effective solutions adapted to this context could be envisaged.

9.4.4. *Definition of consensus algorithms based on artificial intelligence*

The consensus algorithm is one of the elements essential to the operation of blockchain technology. The consensus algorithms used by existing blockchain networks aim essentially to ensure that a malicious node does not seek to add falsified data to blockchain network. To guarantee the operation of the intelligent identity and access management process, it would seem necessary for the reliability of this data to be verified also. Indeed, verifying that the data have not been modified does not make it possible to ensure that it really is correct.

This is why a great deal of research focuses on the idea of coupling AI and consensus mechanisms (Shrestha *et al.* 2018; Yang *et al.* 2019a). Thus, AI techniques in this context, especially fuzzy logic, are used to compare the information added to blockchain register and to ensure that this information is not contradictory and does not contain false information. Nevertheless, this idea has not been applied to identity management. In addition, existing works only propose theoretical solutions which remain to be implemented and evaluated.

9.5. Conclusion

The use of AI techniques to manage identity and access is today strongly supported in the IoT environment. Indeed, this approach could make it possible to use the resources (in terms of calculation, storage and communication) of IoT objects more efficiently and to offer better performances (latency and bandwidth), especially in mobility scenarios. Nevertheless, current solutions for intelligent identity and access management present some limitations, especially then lack of trust, the lack of an efficient load balancing and failure to respect user privacy.

This is why, in this chapter, we have proposed integrating a new technology into the architecture of intelligent identity and access management: blockchain. Indeed, this distributed ledger technology seems to be able to meet the limitations of current solutions. To demonstrate this, in section 9.3, after having introduced an architecture integrating AI and blockchain, we have presented different work using this technology to enable intelligent load calculation, strengthen trust and protect the privacy of users and IoT objects. We have also shown how this technology could enable the advent of new intelligent approaches and guarantee the explainability of the AI process.

However, for a solution for managing identity and access integrating AI and blockchain to see the light of day, some barriers remain to be lifted. Thus, we have presented, in section 9.4, the main challenges: improvement of scaling up, the strengthening of blockchain security, the reduction of energy consumption and the development of new consensus algorithms based on AI.

9.6. References

Abadi, M., Chu, A., Goodfellow, I., McMahan, H.B., Mironov, I., Talwar, K., Zhang, L. (2016). Deep learning with differential privacy. In *2016 ACM SIGSAC Conference on Computer and Communications Security*. ACM, Vienna.

Awais Hassan, M., Habiba, U., Ghani, U., Shoaib, M. (2019). A secure message-passing framework for inter-vehicular communication using blockchain. *International Journal of Distributed Sensor Networks*, 15(2), 1550147719829677.

Bao, X., Su, C., Xiong, Y., Huang, W., Hu, Y. (2019). Flchain: A blockchain for auditable federated learning with trust and incentive. In *2019 5th International Conference on Big Data Computing and Communications (BIGCOM)*. IEEE, Qindgao.

Calvaresi, D., Mualla, Y., Najjar, A., Galland, S., Schumacher, M. (2019). Explainable multi-agent systems through blockchain technology. In *International Workshop on Explainable, Transparent Autonomous Agents and Multi-Agent Systems*. EXTRA AMAS, Montreal.

Chatterjee, B., Das, D., Maity, S., Sen, S. (2018). RF-PUF: Enhancing IoT security through authentication of wireless nodes using in-situ machine learning. *IEEE Internet of Things Journal*, 6(1), 388–398.

Che, Z., Wang, Y., Zhao, J., Qiang, Y., Ma, Y., Liu, J. (2019). A distributed energy trading authentication mechanism based on a consortium blockchain. *Energies*, 12(15), 2878.

Chen, S., Xu, H., Liu, D., Hu, B., Wang, H. (2014). A vision of IoT: Applications, challenges, and opportunities with China perspective. *IEEE Internet of Things*, 1(4), 349–359.

Chen, X., Ji, J., Luo, C., Liao, W., Li, P. (2018). When machine learning meets block-chain: A decentralized, privacy-preserving and secure design. In *2018 IEEE International Conference on Big Data (Big Data)*. IEEE, Seattle.

Cui, Z., Fei, X.U.E., Zhang, S., Cai, X., Cao, Y., Zhang, W., Chen, J. (2020). A hybrid blockchain-based identity authentication scheme for multi-WSN. *IEEE Transactions on Services Computing*, 13(2), 241–251.

Das, R., Gadre, A., Zhang, S., Kumar, S., Moura, J.M. (2018). A deep learning approach to IoT authentication. In *2018 IEEE International Conference on Communications (ICC)*. IEEE, Kansas City.

Dinh, T.N. and Thai, M.T. (2018). AI and blockchain: A disruptive integration. *Computer*, 51(9), 48–53.

Fang, H., Qi, A., Wang, X. (2020a). Fast authentication and progressive authorization in large-scale IoT: How to leverage AI for security enhancement. *IEEE Network*, 34(3), 24–29.

Fang, H., Wang, X., Xu, L. (2020b). Fuzzy learning for multi-dimensional adaptive physical layer authentication: A compact and robust approach. *IEEE Transactions on Wireless Communications*, 19(8), 5420–5432.

Feng, Q., He, D., Zeadally, S., Khan, M.K., Kumar, N. (2019). A survey on privacy protection in blockchain system. *Journal of Network and Computer Applications*, 126, 45–58.

Gervais, A., Karame, G.O., Wüst, K., Glykantzis, V., Ritzdorf, H., Capkun, S. (2016). On the security and performance of proof of work blockchains. In *2016 ACM SIGSAC Conference on Computer and Communications Security*. ACM, Vienna.

Gunning, D. (2017). Explainable artificial intelligence (XAI). Defense Advanced Research Projects Agency (DARPA), 2 [Online]. Available at: https://www.cc.gatech.edu/~alanwags/DLAI2016/(Gunning)%20IJCAI-16%20DLAI%20WS.pdf [Accessed 11 January 2022].

Jameel, F., Javaid, U., Khan, W.U., Aman, M.N., Pervaiz, H., Jäntti, R. (2020). Reinforcement learning in blockchain-enabled IIoT networks: A survey of recent advances and open challenges. *Sustainability*, 12(12), 5161.

Jost, C., Lam, H., Maximov, A., Smeets, B.J. (2015). Encryption performance improvements of the Paillier cryptosystem. *IACR Cryptology ePrint Archive*, 864.

Kang, J., Yu, R., Huang, X., Wu, M., Maharjan, S., Xie, S., Zhang, Y. (2018). Blockchain for secure and efficient data sharing in vehicular edge computing and networks. *IEEE Internet of Things Journal*, 6(3), 4660–4670.

Khalilov, M.C.K. and Levi, A. (2018). A survey on anonymity and privacy in bitcoin-like digital cash systems. *IEEE Communications Surveys & Tutorials*, 20(3), 2543–2585.

Kuo, T.T. and Ohno-Machado, L. (2018). Modelchain: Decentralized privacy-preserving healthcare predictive modeling framework on private blockchain networks [Online]. Available at: https://arxiv.org/abs/1802.01746 [Accessed 11 January 2022].

Laskin, M., Lee, K., Stooke, A., Pinto, L., Abbeel, P., Srinivas, A. (2020). Reinforcement learning with augmented data [Online]. Available at: http://arxiv.org/abs/2004.14990 [Accessed 11 January 2022].

Lin, I.C. and Liao, T.C. (2017). A survey of blockchain security issues and challenges. *IJ Network Security*, 19(5), 653–659.

Liu, C.H., Lin, Q., Wen, S. (2018). Blockchain-enabled data collection and sharing for industrial IoT with deep reinforcement learning. *IEEE Transactions on Industrial Informatics*, 15(6), 3516–3526.

Luu, L., Chu, D.H., Olickel, H., Saxena, P., Hobor, A. (2016). Making smart contracts smarter. In *2016 ACM SIGSAC Conference on Computer and Communications Security*. SIGSAC, Vienna.

McCorry, P., Bakshi, S., Bentov, I., Meiklejohn, S., Miller, A. (2019). Pisa: Arbitration outsourcing for state channels. In *1st ACM Conference on Advances in Financial Technologies*. ACM, Zurich.

Mendiboure, L., Chalouf, M.A., Krief, F. (2018). Towards a blockchain-based SD-IoV for applications authentication and trust management. In *International Conference on Internet of Vehicles*. IEEE, Paris.

Mendiboure, L., Chalouf, M.A., Krief, F. (2020a). Survey on blockchain-based applications in internet of vehicles. *Computers & Electrical Engineering*, 84, 106646.

Mendiboure, L., Chalouf, M.A., Krief, F. (2020b). A scalable blockchain-based approach for authentication and access control in software defined vehicular networks. In *29th International Conference on Computer Communications and Networks (ICCCN 2020)*. IEEE, Honolulu.

Nasir, Q., Qasse, I.A., Abu Talib, M., Nassif, A.B. (2018). Performance analysis of hyperledger fabric platforms. *Security and Communication Networks*, Volume 2018, Article ID 3976093 [Online]. Available at: https://doi.org/10.1155/2018/3976093.

Nassar, M., Salah, K., ur Rehman, M.H., Svetinovic, D. (2020). Blockchain for explainable and trustworthy artificial intelligence. *Wiley Interdisciplinary Reviews: Data Mining and Knowledge Discovery*, 10(1), e1340.

Outchakoucht, A., Hamza, E.S., Leroy, J.P. (2017). Dynamic access control policy based on blockchain and machine learning for the Internet of Things. *International Journal of Advanced Computer Science and Applications*, 8(7), 417–424.

Pandl, K.D., Thiebes, S., Schmidt-Kraepelin, M., Sunyaev, A. (2020). On the convergence of artificial intelligence and distributed ledger technology: A scoping review and future research agenda. *IEEE Access*, 8, 57075–57095.

Punithavathi, P., Geetha, S., Karuppiah, M., Islam, S.H., Hassan, M.M., Choo, K.K.R. (2019). A lightweight machine learning-based authentication framework for smart IoT devices. *Information Sciences*, 484, 255–268.

Qian, Y., Jiang, Y., Chen, J., Zhang, Y., Song, J., Zhou, M., Pustišek, M. (2018). Towards decentralized IoT security enhancement: A blockchain approach. *Computers & Electrical Engineering*, 72, 266–273.

Rathore, S., Pan, Y., Park, J.H. (2019a). BlockDeepNet: A blockchain-based secure deep learning for IoT network. *Sustainability*, 11(14), 3974.

Rathore, S., Kwon, B.W., Park, J.H. (2019b). BlockSecIoTNet: Blockchain-based decentralized security architecture for IoT network. *Journal of Network and Computer Applications*, 143, 167–177.

Salah, K., Rehman, M.H.U., Nizamuddin, N., Al-Fuqaha, A. (2019). Blockchain for AI: Review and open research challenges. *IEEE Access*, 7, 10127–10149.

Shen, M., Tang, X., Zhu, L., Du, X., Guizani, M. (2019). Privacy-preserving support vector machine training over blockchain-based encrypted IoT data in smart cities. *IEEE Internet of Things Journal*, 6(5), 7702–7712.

Shrestha, R., Bajracharya, R., Nam, S.Y. (2018). Blockchain-based message dissemination in VANET. In *2018 IEEE 3rd International Conference on Computing, Communication and Security (ICCCS)*. IEEE, Katmandu.

Singh, S.K., Rathore, S., Park, J.H. (2020). Blockiotintelligence: A blockchain-enabled intelligent IoT architecture with artificial intelligence. *Future Generation Computer Systems*, 110, 721–743.

Xiao, L., Wan, X., Lu, X., Zhang, Y., Wu, D. (2018). IoT security techniques based on machine learning: How do IoT devices use AI to enhance security? *IEEE Signal Processing Magazine*, 35(5), 41–49.

Yang, R., Yu, F.R., Si, P., Yang, Z., Zhang, Y. (2019a). Integrated blockchain and edge computing systems: A survey, some research issues and challenges. *IEEE Communications Surveys & Tutorials*, 21(2), 1508–1532.

Yang, Q., Liu, Y., Chen, T., Tong, Y. (2019b). Federated machine learning: Concept and applications. *ACM Transactions on Intelligent Systems and Technology (TIST)*, 10(2), 1–19.

Zhao, Y., Zhao, J., Jiang, L., Tan, R., Niyato, D., Li, Z., Lyu, L., Liu, Y. (2020). Privacy-preserving blockchain-based federated learning for IoT devices. *IEEE Internet of Things Journal*, 8(3), 1817–1829.

Zhou, J., Cao, Z., Dong, X., Vasilakos, A.V. (2017). Security and privacy for cloud-based IoT: Challenges. *IEEE Communications Magazine*, 55(1), 26–33.

Zhou, S., Huang, H., Chen, W., Zheng, Z., Guo, S. (2019). PiRATE: A blockchain-based secure framework of distributed machine learning in 5G networks. *IEEE Networks*, 34(6), 84–91.

Zhu, X. and Badr, Y. (2018). Identity management systems for the Internet of Things: A survey towards blockchain solutions. *Sensors*, 18(12), 4215.

10

Adapting the Security Level of IoT Applications

Tidiane SYLLA[1], Mohamed-Aymen CHALOUF[2] and Francine KRIEF[3]

[1] *ISA, USTTB, Bamako, Mali*

[2] *IRISA, University of Rennes 1, Lannion, France*

[3] *LaBRI, Bordeaux INP, Talence, France*

10.1. Introduction

The Internet of Things (IoT) refers to the several billion physical devices that are connected to the Internet and able to collect and share data on the real world. These data are processed later to help make decisions. The devices are for the most part formed of objects from daily life (locks, refrigerators, glucometers, etc.). The IoT has been made possible thanks to several factors such as access to low-cost sensor technologies, ubiquitous wireless connectivity (WiFi, 4G/5G, LoRa, etc.), the availability of Cloud/Fog/Edge Computing platforms and artificial intelligence. The number of new connected objects is ever increasing to create new intelligent and evolved services for humans. Thus, the IoT contributes to making users' daily life comfortable and to improving commercial and industrial processes, etc.

Despite the enthusiasm created by the IoT, its adoption is not total. This is due essentially to problems relating to application security, devices and protection of user privacy. Indeed, many attacks have succeeded in compromising the privacy and security of devices and applications. The low capacity of devices and the dynamic nature of the IoT make implementation of classical security and privacy protection solutions poorly adapted. In this regard, it is vital to design and propose security and privacy protection mechanisms adapted to the needs and characteristics of the IoT.

Intelligent Security Management and Control in the IoT,
coordinated by Mohamed-Aymen CHALOUF.

Moreover, a great deal of research has been carried out to provide solutions adapted to security and privacy protection problems in the IoT. One of the possible solutions is the adaptation of the security level of IoT applications depending on the global usage context: user, application, networks, devices, etc.

In this chapter, we begin by recollecting the main characteristics of the IoT (section 10.2). In section 10.3, we present some flagship applications of the IoT. Then, we describe the main architectures of the IoT that have been proposed by standardization bodies (section 10.4). Section 10.5 presents the security and privacy protection services indispensable for securing the IoT. In this section, we also present a state of the art relating to the offer of each of these security services. Then, in section 10.6, we present the idea of guaranteeing a context-aware level of service. This idea consists of adapting the security level of IoT applications depending on the user's global context. Finally, section 10.7 concludes this chapter and presents some perspectives on adapting the security levels of applications in the IoT.

10.2. Definitions and characteristics

10.2.1. *Definitions*

Since the introduction of the term "Internet of Things" by K. Ashton in 1999 (Biron and Follett 2016), the IoT has excited great interest among manufacturers, academics and international standardization bodies. The different visions and technologies of the IoT make its universal definition a difficult task. The International Telecommunications Union (ITU), through its SG20 working group, has defined the IoT in recommendation ITU-T Y.2060 as being "a global infrastructure that interconnects physical and virtual devices using information and communication technologies that make it possible to offer evolved services" (ITU 2012). The International Organization for Standardization (ISO) and the International Electrotechnical Commission (IEC) have, via Joint Technical Committee 1 (JTC1), defined the IoT as "an infrastructure of entities (devices, individuals, systems) and interconnected information resources as well as intelligent services enabling them to process information from the physical and virtual world and to react" (ISO/IEC JTC 1 2014).

10.2.2. *Characteristics*

The characteristics of the IoT (sensitive data, heterogeneity, restricted resources, dynamism, etc.) making the task of securing it specific. To secure this environment, it is necessary to take into account these characteristics and to define an adapted security.

While considering the dynamism of this environment, the best solution, as explained in Sylla *et al.* (2021), is to allow a dynamic and automatic adaptation at security level to be implemented depending on several parameters. These parameters can be regrouped into what we will call the user's global context. In an IoT environment, these parameters can be collected quickly, often thanks to the devices themselves. In addition, it is an environment that facilitates the integration of intelligence in decision-making. This would make it possible to implement dynamic adaptation at security level in IoT applications without needing additional elements. In this section, we present these characteristics.

10.2.2.1. *Sensitive data*

User data collected by IoT devices are for the most part sensitive data. We can for example cite GPS positioning, blood pressure, etc. The data should therefore be processed by protecting user privacy. Indeed, some of this data, such as GPS position and speed, are used to decide the user's global context.

10.2.2.2. *Heterogeneity*

The IoT is formed of devices and systems based on different material, software and network configurations. These devices interact with one another and with different service platforms. This heterogeneity should be considered in IoT applications to enable interoperability and modularity.

10.2.2.3. *Restricted resources*

The low-cost environments in which the devices should be deployed have an impact on their capacities. This translates into restricted resources (processor, RAM, energy and storage).

10.2.2.4. *Dynamism*

IoT applications collect data from the physical world using IoT devices. Dynamic changes in the user's environment can be perceived through the data collected by the devices. This characteristic enables IoT applications to detect and adapt to changes in the physical world. In other words, it makes it possible to determine the user's global context.

10.2.2.5. *Intelligence*

Intelligence is a characteristic of the IoT that consists of implementing knowledge in the IoT by linking algorithms and evolved data processing techniques (machine learning, reasoning, etc.). It helps with dynamic adaptation and decision-making. It enables automatic adaptation of the security level depending on the risks and security parameters linked to the user's context.

10.2.2.6. *Real time*

Real-time data processing is necessary for some critical IoT applications such as those used in patient telemonitoring or even monitoring of forest fires. It is also a need for implementing dynamic adaptation of the security level in IoT applications. Indeed, adaptation of the security level should be carried out continuously and in real time.

10.3. IoT applications

The IoT makes it possible to have a large number of diverse and varied applications such as smart cities, agriculture, logistics and industrial processes.

– *Smart city*: the innovations brought about by implementing the IoT allow cities to become more ecological and durable. Thus, they provide continuous adaptation of these cities' resource usage. A city is called smart if it uses the IoT to improve its efficiency, reduce its costs and improve its citizens' living environment. We distinguish several IoT applications in the smart city: smart homes, e-healthcare, smart water management, smart electricity, smart roads, etc.

– *Intelligent agriculture*: also called smart farming, it consists of putting in place linked modern technologies, especially the IoT, to increase the quality and quantity of agricultural products. Thus, intelligent agriculture makes it possible to have a combination of several technologies, especially GPS, soil humidity analysis, soil quality analysis, etc. Harvested data are used to make decisions related to things such as adapting irrigation or adding fertilizers and/or pesticides.

– *Intelligent factories*: also called a smart factory, it consists of putting in place an integrated and collaborative manufacturing system which in real time meets the variable needs and conditions of the factory, the supply chain and client needs. For example, machines can automatically start production as soon as they receive an order and/or primary materials become available.

10.4. IoT architectures

The many stakeholders in the IoT (industries, international standards and research bodies) have led to many architectures being defined for it. Among these architectures figure the reference architecture from the Industrial Internet Consortium (IIC)[1] and the reference architecture from the ITU (International Telecommunications Union).

The IIC is a consortium of major industrial layers who, in their sectors, are developing information and communication technologies, such as Huawei, Intel, IBM, General Electric, Dell, Microsoft, Ericsson, etc., and some universities. The IIC has proposed a three-tier architecture for industrial IoT systems to enable engineers and architects to design systems from shared concepts and vocabulary (Lin *et al.* 2017). The different tiers are called edge, platform and enterprise. The edge tier manages devices, flow control, access security, etc. The platform tier is responsible for receiving, processing and transmitting control commands (C&C) from the enterprise to edge tiers. The enterprise tier is responsible for implementing applications.

The SG20 working group from the UIT-T has proposed a reference architecture for the IoT (ITU 2012). Unlike that proposed by the IIC, the reference architecture proposed by the UIT-T only targets industrial IoT systems and therefore aims to be more general. This architecture includes four main layers and two transversal layers. These are the sensing layer, the network layer, the service layer and the application layer. The first layer of the architecture is the sensing layer, formed of devices responsible for collecting and transmitting data from the physical world to the upper layers. The network layer manages devices' network connectivity (access control, authentication, mobility management and authorization). The service layer is responsible for the different processings at the service and data stockage layers. The application layer is home to final user applications (smart home, e-health, intelligent agriculture, etc.). The cross-layer management capacity has the role of ensuring the network functions properly. The transversal layer, called the security capacity layer, makes it possible to ensure security services independently of the applications, that is, authentication, access control, integrity, confidentiality, non-repudiation and availability.

10.5. Security, trust and privacy protection in IoT applications

The challenges to be overcome to ensure security, trust and privacy protection in the IoT are significant. In this section, we will define the different mechanisms needed to guarantee these services. We will also introduce the different pieces of research carried out on these mechanisms.

1 https://www.iiconsortium.org/members.htm.

10.5.1. *General remarks*

An IoT environment has a number of security weaknesses, which are consequences of the weakness of various devices and the failure to consider security from the beginning of the design stage (Garcia-Morchon *et al.* 2018). These failures are also linked to poor device settings. So, for example, most devices are issued with factory settings, that is, they are connected to the Internet with the *administrator* identifier by default. With available automatic scanning tools, especially Shodan (Porup 2019), it is then possible even for an individual with little technical skill to find these devices and control them remotely. From a compromised device, the attacker can have access to a larger network. In 2016, the authors of the botnet Mirai created a malicious program able to exploit weaknesses in billions of IP cameras operating using Linux to control them remotely (Woolf 2016). This botnet was, afterward, used to launch a Distributed Denial of Service (DDoS) attack of unusual size against several Internet giants such as Dyn and OVH. The attack succeeded in putting several Dyn servers out of service for more than 10 hours, thus paralyzing Dyn users: Twitter, Paypal, Netflix, etc. In December 2019, the researchers Bitdefender discovered a malicious program much more powerful and resilient than Mirai, bearing the name Dark Nexus (Arsene 2020). This botnet is able to execute DDoS type attacks, as well as malicious programs on the compromised devices.

User privacy is impacted by the use of IoT applications (Zhang *et al.* 2017). IoT devices collect a large amount of users's data without their explicit consent, which may contain sensitive information. However, the more data there is on a person, the more possible it is to cross-check this data, infer from it and establish a profile of the person. The consequence of this is that it is possible to carry out mass espionage on users of these applications. The risks for privacy are also linked to the implementation of integrity and confidentiality of personal data when it is collected, processed and stored. Indeed, when these security services are not correctly assured or if they are compromised, the data can be exploited for a number of malicious ends: sale on the Darknet (Patterson and Kates 2019), espionage, etc. Finally, protection of privacy has a regulatory aspect, that is, it is regulated by texts (laws) controlled by politicians. These texts consider socio-economic, cultural and religious factors that can vary from one country to another. Thus, this variable aspect of privacy protection should be considered in the development of IoT applications.

10.5.2. *Security services*

In this section, we discuss the security services needed for the proper functioning of IoT applications and devices. For each service, we summarize the definition and analyze the different pieces of research impacting the implementation of mechanisms making it possible to ensure this IoT service.

10.5.2.1. *Authentication and access control*

10.5.2.1.1. Authentication

Definition

Authentication consists of verifying the authenticity of an entity's identity with the aim of authorizing it to access a protected resource. It includes identifying an entity using information called an identifier. An identifier is a piece of information that identifies an entity formally, uniquely and without ambiguity: for example, social security number and national identity number. Identification is the process preceding authentication.

Research

Authentication is situated at three levels in the IoT: user, devices and messages exchanged. User authentication depends strongly on the type of IoT application. Indeed, some IoT applications are accessible via applications installed on a smartphone/computer, whereas others are accessible via rudimentary man-machine interfaces. Therefore, research should be carried out to allow secure and unfalsifiable authentication of the user, whatever type of IoT application is involved. Several pieces of research have proved the effectiveness of multi-factor authentication in Web/mobile applications (Kambou and Bouabdallah 2019). It is possible to link this authentication method with the dynamic character of the IoT, taking account of the user's environment at the moment they access the application. This adaptation takes into account relevant contextual information (see section 10.6.1).

Research on the authentication of devices in the IoT is sub-divided into two large axes. The first concerns lightweight cryptographic protocols for constrained IoT devices. On this axis, several pieces of research have been carried out, of which most focused on Elliptical Curve Cryptography (ECC). This interest is explained mainly by the very good security properties offered by the ECC for constrained devices. Indeed, the implementation of ECC consumes very little energy and memory, unlike other cryptographic algorithms (Goyal and Sahula 2016; Dubois 2018). The second axis focuses on unique physical device identifiers called physically unclonable functions (PUFs) associated with cryptography algorithms (ECC) to allow for a safe and reliable authentication (Akhundov *et al.* 2019).

10.5.2.1.2. Access control

Definition

Access control aims to verify the validity of an authorization for access to a protected resource. It makes it possible to ensure that only authenticated and authorized entities can access a protected resource and contribute to protecting user privacy. It is generally implemented by a server that receives the access requests to a resource and grants or refuses access to the requesting entity according to previously defined authorizations. Implementation of access control in the IoT requires adaptation of traditional access control mechanisms.

Research

Traditional access control mechanisms used, such as Access Control Lists (ACL) or Role-Based Access Control (RBAC), are not directly applicable in IoT applications. This is linked to the characteristics of IoT devices and applications on the one hand, and to the complexity, absence of flexibility and support for heterogeneity in these mechanisms on the other hand. These research activities focus on Attribute-Based Access Control (ABAC) and Contextual Role-Based Access Control (CRBAC). These mechanisms are flexible and evolutive, hence their suitability for the IoT.

CRBAC is based on the definition of roles linked to the users' different overall contexts to permit access to resources to an entity possessing the appropriate roles and existing in the associated overall context (Kulkarni and Tripathi 2008). ABAC is a system in which access rights are attributed to users by a combination of attributes (e.g. localization, access tokens, identifiers, post occupied and PIN code). Several pieces of research have been carried out to find ABAC mechanisms for the IoT. Thus, in Yao *et al.* (2015), a lightweight ABAC system for the IoT has been proposed. This system is based on using the ECC and has the objective of solving security and privacy protection problems in the IoT. In Oualha and Nguyen (2016), the authors adapted Ciphertext Policy Attribute-Based Encryption (CP-ABE) (Bethencourt *et al.* 2007) for the IoT. CP-ABE was not originally adapted to the IoT because it is cumbersome. Despite this research, a great deal remains to be done in this domain. Research can be carried out for ABAC systems, of limited duration and based on the use of contextual access tokens. In this direction, in Alphand *et al.* (2018), the authors propose an authorization management based on the use of access tokens resulting from the authentication and authorization architecture for a constrained environment (CEA) (Shaad *et al.* 2018). This system can be adapted with the help of contextual access tokens. In addition to RBAC and ABAC systems, research focuses on other access control systems for the IoT. Thus, the Risk-Based Access Control (RiBAC) system evaluates risk when authorizing an access request to grant or refuse access (Atlam *et al.* 2018). We also distinguish Identity Authentication and Capability Based Access Control (IACAC).

10.5.3. *Communication security*

Communication security consists of securing the different communication channels against eavesdropping, data alteration, the intelligibility of messages transmitted by unauthorized agents, etc.

10.5.3.1. *Integrity*

10.5.3.1.1. Definition

Data integrity consists of ensuring that the data have not undergone any deliberate or accidental modification during transmission, processing and storage. In practice, data integrity is implemented by cryptographic hashing, an algorithm implementing a one-way mathematical function. This algorithm takes variable amounts of input data and produces a string of characters of fixed size called a digest. Authenticating the origin of a piece of data consists of ensuring the authenticity of the data received.

10.5.3.1.2. Research

Data integrity is generally associated with authentication of the data's origin. Both these services make it possible to evade attacks against data modification targeting IoT applications. Several pieces of research and standardization have given interesting results on these points. For example, variants of the Secure Hash Algorithm (SHA) make it possible to generate unique digest whose collision rate is almost null (Handschuh 2005). There are six hash functions with different security levels: SHA-0, SHA-1, SHA-224, SHA-256, SHA-384 and SHA-512. All these functions are not applicable to IoT devices for reasons of capacity. Nevertheless, hardware IoT solutions are now available to allow material acceleration of some variants such as SHA-1. Thus, research should be carried out to permit the hardware acceleration of variants with a higher security level, such as SHA-256 or SHA-512, for the IoT and resistant to quantum computers. Indeed, SHA-1 and SHA-0 are not secure (Seals 2020).

Key hash message authentication algorithms (HMAC) are used instead: HMAC-SHA-1, HMAC-SHA-256 and HMAC-SHA-512. Research should be carried out to permit the hardware integration of HMAC algorithms with a high security device into IoT devices. Research should also be carried out to implement data integrity via the blockchain. For example, in Javaid *et al.* (2018), the authors have implemented a system making it possible to verify the authenticity of data using PUF and the blockchain. However, since the blockchain requires a high computational capacity, energy and latency, proxies with very low latency should be put in place. Indeed, the need for proxies is explained by the fact that the vast majority of IoT devices cannot directly communicate with a blockchain.

10.5.3.2. *Confidentiality*

10.5.3.2.1. Definition

Confidentiality makes it possible to render data intelligible only for the intended recipients. To ensure the communications security, data confidentiality should be assured during transfer and processing. It is implemented through use of cryptography algorithms. A key is used to encrypt and decrypt the data. According to Kerckhoffs' principle (Kerckhoffs 1883), the power of a cryptography system lies in the secret keys. Thus, depending on the key management mode, we distinguish two branches of cryptographic systems: symmetrical key cryptography and asymmetrical key cryptography.

10.5.3.2.2. Research

Confidentiality in IoT applications has been the subject of many pieces of research. In this sense, several algorithms have been developed for IoT devices. Moreover, some existing algorithms have also been adapted for the IoT with the reduction of the number of operations necessary for encryption/decryption operations and the reduction of key size. The advanced standard cryptographic encryption system (AES) is the most used. Work carried out in Singh *et al.* (2017) presents a state of the art on lightweight and adapted cryptographic primitives with a security level sufficient for the IoT.

The advent of quantum computing has broken the promises made for robust and known cryptographic systems such as ECC and AES. Because of this, research should be carried out to find new cryptographic systems for IoT devices. Interesting research pathways target systems called post-quantum cryptographic systems or Post-Quantum Cryptography (Chen *et al.* 2016b). Thus, in Cheng *et al.* (2017), the authors present the risks that quantum computers represent for current cryptographic systems and give indications for designing resistant systems in the quantum age. One of the possible pathways is the use of the cryptographic system NTRU (N-th degree Truncated polynomial Ring Units) (Cheng *et al.* 2020) created in 1996 by Jeffrey Hoffstein, Jill Pipher and Joseph H. Silverman.

10.5.4. *Trust*

10.5.4.1. *Definition*

Trust consists of ensuring that the characteristics an entity indicates about its capacity to fulfill certain tasks and its properties really are what it claims they are and that it can correctly fulfill these tasks. It also consists of ensuring the authenticity of the entity's identity. Trust management is a complex process.

10.5.4.2. *Research*

Trust in the IoT has been the object of much research. In Chen *et al.* (2019), the authors present an architecture of trust for the IoT based on reputation and relying on an SDN (Software Defined Network). This architecture is coupled with an authorization management system. In Chen *et al.* (2016a), the authors proposed a trust management system for the IoT based on SOA (Service Oriented Architecture) models and Social-IoT. The proposed solution uses a distributed collaborative filtering for selecting trusted feedback. A filter can be the similarity ranking for friendships.

Research has also been carried out on trust in edge-level processing. Trust in the processing can be assured by an element called a Root of Trust (RoC). This element, secured at hardware level, enables management of random numbers, private keys and session keys needed for secure encryption. Many devices are equipped with RoC. The design of RoC modules follows the trusted Platform Module Standard (PMS) from ISO/CEI (ISO/JTC 1 2015). They are sometimes mounted on the secure area of a processor, still called a Trusted Execution Environment (TEE). TEE is a hardware security module that allows provision of a code execution environment isolated in the processor. Thus, it enables a high level of trust in the code executed within it, especially the sensitive part of applications: generation of random numbers, key derivation, powerful calculation, etc. In industry, it is present on some Intel and ARM processors, under the names Intel SGX and ARM TrustZone (Li *et al.* 2019). Thus, research can be carried out to integrate TEEs into IoT devices for more hardware security (Garcia-Morchon *et al.* 2018).

Work is also being carried out to facilitate integration of the blockchain as a trusted support in processing in the IoT. Research carried out in Hammi *et al.* (2018) presents a decentralized trust management system allowing authentication and authorization management of IoT devices. In Dedeoglu *et al.* (2019), the authors have proposed an end-to-end trust management architecture for the IoT based on the blockchain.

10.5.5. *Privacy*

10.5.5.1. *Definition*

Privacy protection consists of implementing all the techniques and methods necessary to ensure that users' personal data are not stolen or divulged without their consent. In other words, privacy protection allows the user to have control over their data.

10.5.5.2. *Research*

The results of research carried out up until now have shown that it is possible to preserve the privacy of users in the IoT. Many of these results have their advantages and disadvantages (Abi Sen *et al.* 2018). However, despite the extent of work to date, it must be stated that no standard privacy protection mechanism has been created similar to secure communication protocols such as the Secure Socket Layer/Transport Layer Security (SSL/TLS) or cryptographic algorithms such as AES and ECC. A standard mechanism should be able to allow actors evolving in the IoT to put in place privacy protection measures in different IoT applications. Thus, work has been carried out for a standard privacy protection mechanism.

Research should also be carried out on user-centric privacy protection in the IoT (Barhamgi *et al.* 2018). This approach considers non-technical aspects (see section 10.5.5.1) in addition to technical aspects of privacy protection. Moreover, it is easier for users to comprehend. This is why in Torre *et al.* (2016), the authors proposed a framework for the protection of personal data centered on the user in the IoT. The limitation of this framework is that it is static. In Barhamgi *et al.* (2018), the authors proposed an architecture that allows the user to know what risks they are incurring when they share their data in a smart home. This architecture is adaptive to user context.

10.6. Adapting the security level in the IoT

Several pieces of research have been carried out to provide security and privacy protection mechanisms adapted to IoT applications. The solutions proposed generally target the proposition and improvement of security, trust or privacy protection mechanisms, either specific ones or a combination of mechanisms. These mechanisms are proposed to meet a specific threat model to which the target system is exposed. Indeed, the mobility of IoT application users makes these static mechanisms limited, since the threat models to which these applications are exposed are dynamic.

Adapting the security level of IoT applications makes it possible to overcome the challenge of continuous and dynamic security and privacy protection for applications. It consists of evaluating dynamically and intelligently the threats and risks linked to a user context. The implementation of security and privacy protection mechanisms is then adapted according to the threats and risks present in this context. Intelligence consists of using the data collected by the devices to determine the user's current situation and the risks present within it. For example, in a low-risk situation, a classic security mechanism will still maintain the same level, unlike the adaptive security mechanism that will lower the security level and facilitate access to the application. In the following sections, we will recap on the definition of context-awareness in the IoT and present notions relating to adaptation of security levels in IoT applications.

10.6.1. *Context-awareness*

Context-awareness was mentioned for the first time by Schilit and Theimer in 1994. It is based on the notion of context (Abowd *et al.* 1999), which is defined in the following way: "The context is any information that can be used to characterize the situation of an entity, since an entity can be a person, a place, or a physical or IT object" (Abowd *et al.* 1999). A great deal of information is necessary to identify a context with precision. According to Abowd and Mynatt (2000), the questions that must be answered to identify a context are: who, what, where, when and why. This information is called contextual information. The process of acquiring it is facilitated by sensors and other IoT devices. As contextual information, we can cite geographic position, speed of movement, etc.

Context-aware computing has been defined by Abowd *et al.* in 1999 as follows: "A system is context-aware if it uses context to provide relevant information and/or services to the user, where the relevance refers to the task being carried out by the user" (Abowd *et al.* 1999). From this definition and the intrinsic characteristics of the IoT, Sylla *et al.* (2021) have defined a context-aware IoT system as being a system that implements the context to acquire appropriate and relevant information to optimize services through dynamic actions and adaptations carried out without the user's intervention.

Moreover, context-aware computing relies on models of contexts and context attributes (Perera *et al.* 2014). A context model is a sub-set of the context accessible from contextual information. A context attribute is an element of the context model that makes it possible to describe the context. Another fundamental notion of context-sensitivity is Quality of Context (QoC). According to Perera *et al.* (2014), "QoC is defined with the help of a set of parameters that expresses the quality of the requirements and properties of contextual data". Thus, QoC improves context detection quality by reducing errors due to sensor faults and approximations.

A context has a lifecycle formed of four stages (Perera *et al.* 2014). Figure 10.1 illustrates the different stages of this cycle, which are acquisition, modeling, reasoning and dissemination. Acquisition consists of collecting raw contextual information from devices called context sources. We distinguish several acquisition techniques: request/response, regular frequency, pub/sub, etc. Modeling consists of processing contextual information received in the form of attributes, characteristics and QoC.

We distinguish several context-modeling techniques: thing-oriented, key-value, ontology, logic, etc. Reasoning uses the information resulting from the modeling to determine the context. The context resulting from this stage can be disseminated to consumers of the context. We distinguish several reasoning techniques: rules, decision-trees, Naive Bayes, fuzzy logic, etc.

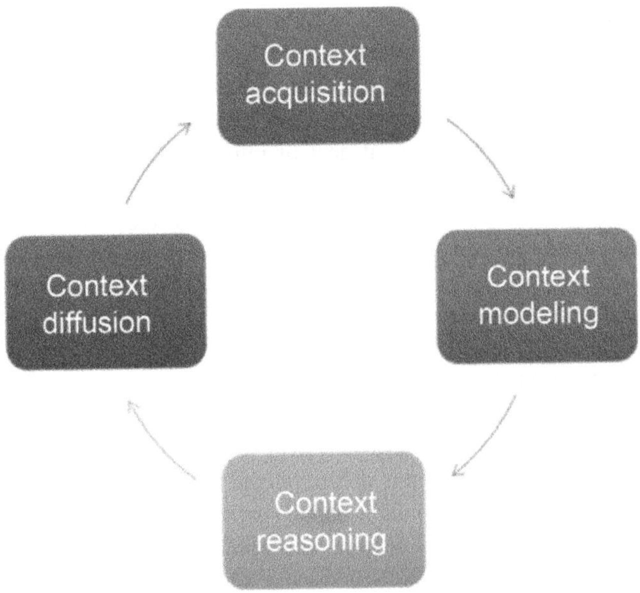

Figure 10.1. *Lifecycle of a context. For a color version of this figure, see www.iste.co.uk/chalouf/intelligent.zip*

10.6.2. *Context-aware security*

Context-aware security, or adapting the security level depending on context, is a branch of IT security that has appeared recently with the emphasis on IT technologies.

Context-aware security in the IoT can be defined as being the application of security and privacy protection mechanisms depending on the user's situation (and depending on the risks and threats present) without their explicit intervention (Sylla *et al.* 2021). Indeed, in IoT applications, the user's context changes quite often. Thus, a security level sufficient at a given instant t may not be sufficient at instant $t + 1$. The need for adaptive security in the IoT is intrinsically linked to device protection requirements, personal data and applications. Figure 10.2 illustrates an example of implementing context-aware security in the IoT.

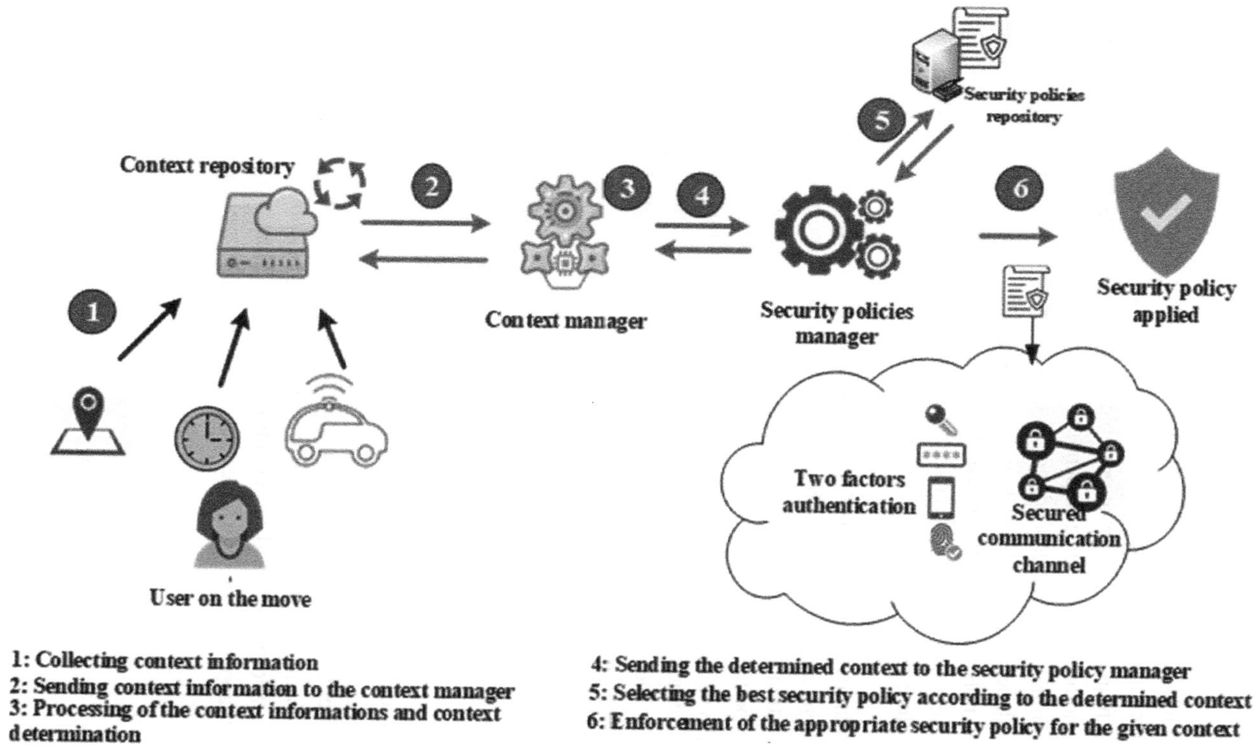

Figure 10.2. *Example of implementing context-aware security (Sylla et al. 2021). For a color version of this figure, see www.iste.co.uk/chalouf/intelligent.zip*

Context-aware authentication allows us to strengthen or lessen authentication by choosing between a two-factor or single-factor authentication.

For example, when Alice visits her doctor at the hospital, the doctor can use simple authentication to access Alice's data. On the other hand, when Alice is at home, a two-factor authentication will be deployed to strengthen the security level. Context-aware authentication management makes it possible to control an entity's access to the user's applications and/or personal data depending on the entity's rights and context. The security of communications depending on the context makes it possible to choose dynamically, depending on the network and security risks, to establish a secure connection or not. For example, when Alice wants to connect to her home control application using public WiFi, considering the risks, a secure connection is first established before any data are exchanged.

Context-adaptive privacy protection consists of implementing privacy protection measures depending on the user's context and the risks associated with this context. Privacy protection measures translate into context-aware privacy protection policies. One or more policies can be applied for a single context. For example, for Alice, a policy can specify complete anonymization and data encryption before sending data to the hospital's epidemiology application in the morning and on the doctor's request, without which the data will not be sent. This adaptation of security level makes it difficult to violate privacy, even in cases of user negligence.

10.6.3. *Context-aware security architecture and privacy protection designed using the "as a service" approach*

In general, existing solutions for adapting the security level of IoT applications address only the implementation of a single security or privacy protection mechanism.

Furthermore, these solutions target a specific application. Indeed, the user's overall context can present multiple security threats. Consequently, it can necessitate applying a set of security and privacy and protection mechanisms. Moreover, a user may have a variety of IoT applications. Thus, one specific security per application is not a viable model.

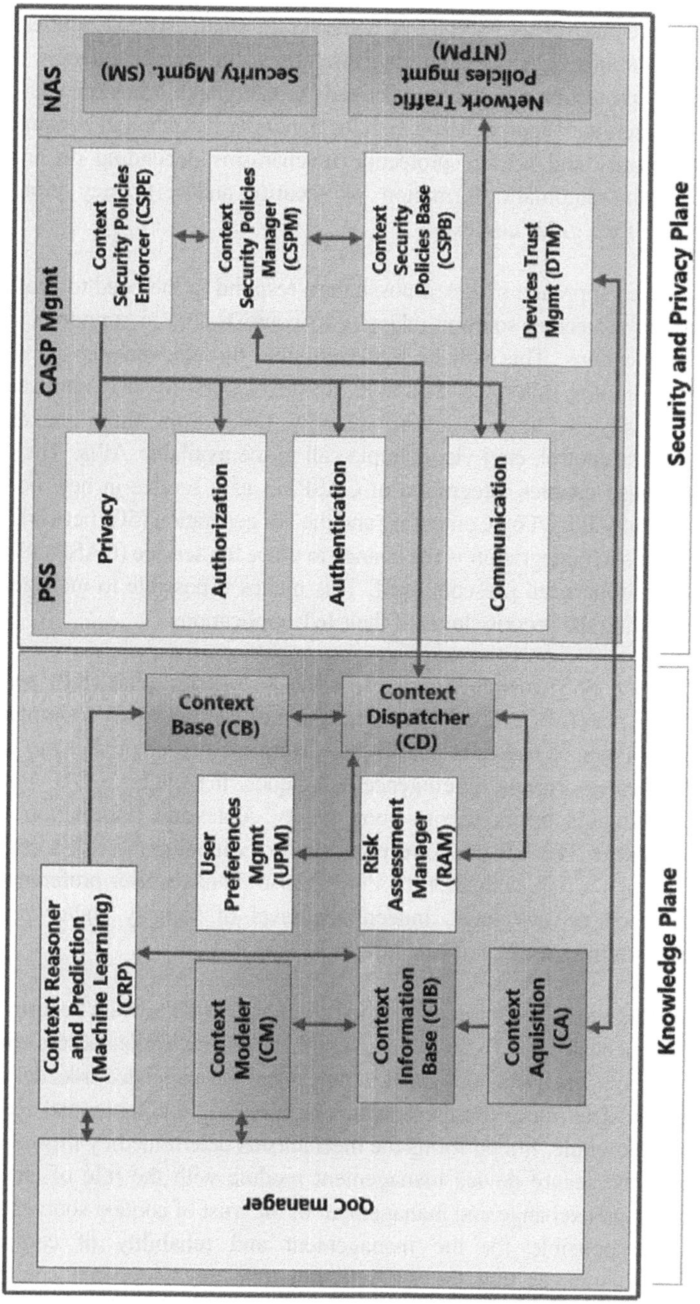

Figure 10.3. *CASPaaS architecture (Sylla et al. 2020). For a color version of this figure, see www.iste.co.uk/chalouf/intelligent.zip*

To this effect, in Sylla *et al.* (2020), we have proposed a context-aware security and privacy protection architecture as a service (CASPaaS) for smart city IoT applications. Unlike existing solutions, CASPaaS architecture aims for implementation of all security and privacy protection mechanisms depending on context. It has been designed following the "as a service" approach and enables dynamic, flexible and personalized enforcement of security and privacy protection mechanisms depending on context. CASPaaS allows the automatic formation of security and/or privacy protection mechanisms needed for a given context.

The "as a service" approach makes it possible to respond to the need to provide a security and privacy protection solution adaptive to context, able to support a single user's various applications. This will be made possible through making the secure Application Programming Interface available to developers of IoT applications. Developers will easily be able to export security tasks from their applications (authentication, access control, etc.) via a simple call to the available APIs. The "as a service" approach also enables integration of CASPaaS as a service in new network architectures, especially Edge/Fog Computing and the 5th generation (5G) network. One of the advantages of such integration is the chance to place the service (CASPaaS) near to the user each time the need presents itself. This makes it possible to manage user mobility when adapting the security level of their IoT applications.

CASPaaS (Figure 10.3) includes two plans: the knowledge plan (KP) and the security and privacy plan (SPP). The KP includes the modules necessary to implement intelligence in CASPaaS. It makes it possible to determine and/or predict the user's overall context by using artificial intelligence techniques. It includes a QoC module with the role of avoiding errors due to poor quality contextual information when determining the context. The KP is also responsible for evaluating the risk associated with a context through a risk evaluation module. It also manages user preferences in terms of security and privacy level. Indeed, the level of security and/or privacy protection depends on the specifics of each user.

The SPP includes modules (services) needed to apply context-aware security and privacy protection mechanisms. These are, among others, authentication, access control, communication security, etc. Depending on the context and the associated risks delivered by the KP, this plan determines the mechanisms required via the contextual security policy management module. It then forms the mechanisms determined by this module. The SPP includes the secure device management module with the role of ensuring security of information exchange and management of the trust of context sources. This module is also responsible for the management and reliability of contextual information. This guarantees that the KP will not treat any unreliable contextual information or any information from a malicious device. The SPP also includes the security management module, which has the role of ensuring the security of the architecture against attacks, especially denial of service attacks. The network traffic policy management module makes it possible to send the security and privacy protection measures to be applied by the user's devices.

CASPaaS integrates perfectly into the reference architecture from the ITU-T for the IoT. It shows several advantages. First, the separation of intelligence (decision-making) and the application of security and privacy protection mechanisms make it possible to have a flexible and evolutive solution. Second, the "as a service" design enables support of horizontal applications and the integration of network-oriented services. Third, it can be deployed on the fly as a service on an edge network.

10.7. Conclusion

The development and adoption of IoT applications requires that security, trust and privacy protection must be guaranteed. To meet this challenge, research emphasizes the importance of adapting the security of IoT applications. This adaptation should be made in a secure and trusted environment to avoid any compromise. Several security and privacy protection mechanisms are needed to meet the threats and risks associated with a context. Thus, work on implementing *security level adaptation* in applications should consider the security of context-aware management as well as support for all security and privacy protection mechanisms. Moreover, this work should also consider mutualizing applications' security and privacy protection mechanisms. This will contribute to balancing the security–ease of use equation.

Furthermore, *privacy* protection in IoT applications used in citizens' daily life includes several aspects. These aspects, especially respect for judicial regulations in effect and supporting user preferences, are vital and should be integrated into adaptive security solutions. On the other hand, the integration of non-technical aspects will allow, in addition to adaptation of the security level depending on the context, implementation of the user-centric approach. In fact, this approach contributes to a safer adoption and development of IoT applications by populations.

10.8. References

Abi Sen, A.A., Eassa, F.A., Jambi, K., Yamin, M. (2018). Preserving privacy in Internet of Things: A survey. *International Journal of Information Technology*, 10(2), 189–200.

Abowd, G.D. and Mynatt, E.D. (2000). Charting past, present, and future research in ubiquitous computing. *ACM Transactions on Computer-Human Interaction*, 7(1), 29–58.

Abowd, G.D., Dey, A.K., Brown, P.J., Davies, N., Smith, M., Steggles, P. (1999). Towards a better understanding of context and context-awareness. In *1st International Symposium on Handheld and Ubiquitous Computing*. HUC, Karlsruhe.

Akhundov, H., van der Sluis, E., Hamdioui, S., Taouil, M. (2019). Public-key based authentication architecture for IoT devices using PUF. In *6th International Conference on Computer Science, Engineering and Information Technology*. CSEIT, Zurich.

Alphand, O., Amoretti, M., Claeys, T., Dall'Asta, S., Duda, A., Ferrari, G., Rousseau, F., Tourancheau, B., Veltri, L., Zanichelli, F. (2018). IoTChain: A blockchain security architecture for the Internet of Things. In *2018 IEEE Wireless Communications and Networking Conference (WCNC)*. IEEE, Barcelona.

Arsene, L. (2020). New dark_nexus IoT Botnet puts others to shame [Online]. Available at: https://labs.bitdefender.com/2020/04/new-dark_nexus-iot-botnet-puts-others-to-shame/ [Accessed 10 June 2020].

Atlam, H.F., Alenezi, A., Hussein, R.K., Wills, G. (2018). Validation of an adaptive risk-based access control model for the Internet of Things. *International Journal of Computer Network and Information Security*, 10(1), 26–35.

Aubonnet, T., Amina, B., Lemoine, F., Simoni, N. (2016). Controlled components for Internet of Things as-a-service. *Open Journal of Internet of Things (OJIOT)*, 2(1), 16–33.

Barhamgi, M., Perera, C., Ghedira, C., Benslimane, D. (2018). User-centric privacy engineering for the Internet of Things [Online]. Available at: https://arxiv.org/abs/1809.00926 [Accessed 15 June 2020].

Bethencourt, J., Sahai, A., Waters, B. (2007). Ciphertext-policy attribute-based encryption. In *2007 IEEE Symposium on Security and Privacy*. IEEE, Berkeley.

Biron, J. and Follett, J. (2016). *Foundational Elements of an IoT Solution*, 1st edition. O'Reilly Media, Inc., Newton.

Chen, I.-R., Guo, J., Bao, F. (2016a). Trust management for SOA-based IoT and its application to service composition. *IEEE Transactions on Services Computing*, 9(3), 482–495.

Chen, L., Jordan, S., Liu, Y.-K., Moody, D., Peralta, R., Perlner, R., Smith-Tone, D. (2016b). Report on post-quantum cryptography. Report, US Department of Commerce, National Institute of Standards and Technology.

Chen, J., Tian, Z., Cui, X., Yin, L., Wang, X. (2019). Trust architecture and reputation evaluation for Internet of Things. *Journal of Ambient Intelligence and Humanized Computing*, 10(8), 3099–3107.

Cheng, C., Lu, R., Petzoldt, A., Takagi, T. (2017). Securing the Internet of Things in a quantum world. *IEEE Communications Magazine*, 55(2), 116–120.

Cheng, H., Dinu, D., Großschädl, J., Rønne, P.B., Ryan, P.Y.A. (2020). A lightweight implementation of NTRU prime for the post-quantum Internet of Things. In *Information Security Theory and Practice*, Laurent, M., Giannetsos, T. (eds). Springer International Publishing, Cham.

Dedeoglu, V., Jurdak, R., Putra, G.D., Dorri, A., Kanhere, S.S. (2019). A trust architecture for blockchain in IoT. In *16th EAI International Conference on Mobile and Ubiquitous Systems: Computing, Networking and Services. MobiQuitous: Computing, Networking and Services*. ACM, Houston.

Dubois, C. (2018). MIT's encryption chip reduces public-key encryption power consumption by 99.75% [Online]. Available at: https://www.allaboutcircuits.com/news/energy-efficient-public-key-encryption-chip-solve-iot-security-problems/ [Accessed 23 September 2018].

Garcia-Morchon, O., Kumar, S., Sethi, M. (2018). State-of-the-art and challenges for the Internet of Things security [Online]. Available at: https://tools.ietf.org/id/draft-irtf-t2trg-iot-seccons-13.html [Accessed 7 February 2019].

Goyal, T.K. and Sahula, V. (2016). Lightweight security algorithm for low power IoT devices. In *2016 International Conference on Advances in Computing, Communications and Informatics (ICACCI)*. IEEE, Jaipur.

Hammi, M.T., Hammi, B., Bellot, P., Serhrouchni, A. (2018). Bubbles of trust: A decentralized blockchain-based authentication system for IoT. *Computers & Security*, 78, 126–142.

Handschuh, H. (2005). SHA family (Secure Hash Algorithm). In *Encyclopedia of Cryptography and Security*, Tilborg, H.C.A. (ed.). Springer, New York.

ISO/IEC JTC 1 (2014). Internet of Things (IoT), preliminary report [Online]. Available at: https://www.iso.org/files/live/sites/isoorg/files/developing_standards/docs/en/internet_of_things_report-jtc1.pdf [Accessed 2 August 2019].

ISO/JTC 1 (2015). ISO/IEC 11889-1:2015 [Online]. Available at: https://www.iso.org/cms/render/live/en/sites/isoorg/contents/data/standard/06/65/66510.html [Accessed 17 June 2020].

ITU (2012). ITU-T Recommendation Y.4000/Y.2060, June [Online]. Available at: https://www.itu.int/ITU-T/recommendations/rec.aspx?rec=y.2060 [Accessed 2 July 2019].

Javaid, U., Aman, M.N., Sikdar, B. (2018). BlockPro: Blockchain based data provenance and integrity for secure IoT environments. In *1st Workshop on Blockchain Enabled Networked Sensor Systems*. ACM, Shenzhen.

Kambou, S. and Bouabdallah, A. (2019). A strong authentication method for web/mobile services. In *2019 6th IEEE International Conference on Cyber Security and Cloud Computing (CSCloud)/2019 5th IEEE International Conference on Edge Computing and Scalable Cloud (EdgeCom)*. IEEE, Paris.

Kerckhoffs, A. (1883). La cryptographie militaire. *Journal des Sciences Militaires*, 11, 161–191.

Kulkarni, D. and Tripathi, A. (2008). Context-aware role-based access control in pervasive computing systems. In *13th ACM Symposium on Access Control Models and Technologies*. ACM, Estes Park.

Li, W., Chen, H., Chen, H. (2019). Research on ARM TrustZone. *GetMobile: Mobile Computing and Communications*, 22(3), 17–22.

Lin, S.-W., Miller, B., Durand, J., Bleakley, G., Chigani, A., Martin, R., Murphy, B., Crawford, M. (2017). The industrial Internet of Things volume G1: Reference architecture. Report, IIC [Online]. Available at: https://www.iiconsortium.org/IIC_PUB_G1_V1.80_2017-01-31.pdf [Accessed 19 February 2019].

Oualha, N. and Nguyen, K.T. (2016). Lightweight attribute-based encryption for the Internet of Things. In *2016 25th International Conference on Computer Communication and Networks (ICCCN)*. IEEE, Waikoloa.

Patterson, D. and Kates, G. (2019). We found our personal data on the dark web. Is yours there, too? [Online]. Available at: https://www.cbsnews.com/news/we-found-our-personal-data-on-the-dark-web-is-yours-there-too/ [Accessed 12 June 2020].

Perera, C., Zaslavsky, A., Christen, P., Georgakopoulos, D. (2014). Context aware computing for the Internet of Things: A survey. *IEEE Communications Surveys & Tutorials*, 16(1), 414–454.

Porup, J.M. (2019). What is Shodan? The search engine for everything on the internet [Online]. Available at: https://www.csoonline.com/article/3276660/what-is-shodan-the-search-engine-for-everything-on-the-internet.html [Accessed 10 June 2020].

Ramos, J.L.H., Bernabe, J.B., Skarmeta, A.F. (2015). Managing context information for adaptive security in IoT environments. In *2015 IEEE 29th International Conference on Advanced Information Networking and Applications Workshops*. IEEE, Gwangju.

Seals, T. (2020). Exploit fully breaks SHA-1, lowers the attack bar [Online]. Available at: https://threatpost.com/exploit-fully-breaks-sha-1/151697/ [Accessed 15 June 2020].

Shaad, J., Danyliw, R., Kaduk, B. (2018). Authentication and authorization for constrained environments (ace) [Online]. Available at: https://datatracker.ietf.org/wg/ace/about/ [Accessed 11 December 2018].

Singh, S., Sharma, P.K., Moon, S.Y., Park, J.H. (2017). Advanced lightweight encryption algorithms for IoT devices: Survey, challenges and solutions. *Journal of Ambient Intelligence and Humanized Computing*, 1–18.

Sylla, T., Chalouf, M.A., Krief, F., Samaké, K. (2020). Towards a context-aware security and privacy as a service in the Internet of Things. In *Information Security Theory and Practice*, Laurent, M., Giannetsos, T. (eds). Springer International Publishing, Cham.

Sylla, T., Chalouf, M.A., Krief, F., Samaké, K. (2021). Context-aware security in the Internet of Things: A survey. *International Journal of Autonomous and Adaptive Communications Systems*, 14(3), 231–263.

Torre, I., Koceva, F., Sanchez, O.R., Adorni, G. (2016). A framework for personal data protection in the IoT. In *2016 11th International Conference for Internet Technology and Secured Transactions (ICITST)*. IEEE, Barcelona.

Woolf, N. (2016). DDoS attack that disrupted internet was largest of its kind in history, experts say. 26 October 2016 [Online]. Available at: https://www.theguardian.com/technology/2016/oct/26/ddos-attack-dyn-mirai-botnet [Accessed 4 November 2018].

Yao, X., Chen, Z., Tian, Y. (2015). A lightweight attribute-based encryption scheme for the Internet of Things. *Future Generation Computer Systems*, 49, 104–112.

Zhang, K., Ni, J., Yang, K., Liang, X., Ren, J., Shen, X.S. (2017). Security and privacy in smart city applications: Challenges and solutions. *IEEE Communications Magazine*, 55(1), 122–129.

11

Moving Target Defense Techniques for the IoT

Renzo E. NAVAS, Laurent TOUTAIN and Georgios Z. PAPADOPOULOS

IRISA, IMT Atlantique, Rennes, France

This chapter addresses cyberdefense techniques for the Internet of Things (IoT) based on the Moving Target Defense (MTD) paradigm. The MTD paradigm is intended to balance the current asymmetry – in terms of information – between a static system and unknown attackers. In a static system context, an attacker potentially has infinite time to find and exploit weaknesses in the system. To limit opportunities for attackers, MTD proposes actively dynamizing (previously static) resources. Indeed, in this dynamic system context, the validity of the information the attacker learns is now time-limited. *A priori*, this dynamism through design means the attacker needs to make more effort to succeed. Even though the use and development of MTD techniques is extensive in the bibliography, techniques adapted to the IoT remain limited.

In the course of this chapter, we will present the concepts linked to MTD in more detail, explore existing MTD techniques by prioritizing more particularly those that are designed to consider the specific constraints of the IoT, evaluate the feasibility of MTD as a cyberdefense tool for IoT systems-particularly in the domain of network layers – and present a generic *framework* to help with designing and implementing MTD techniques adapted to the IoT.

Intelligent Security Management and Control in the IoT,
coordinated by Mohamed-Aymen CHALOUF.

11.1. Introduction

The IoT is formed of billions of connected devices that interact with the physical world. It lies at the boundary of the digital and physical domains, and has applications in many aspects of modern societies such as agriculture, industrial automation (industry 4.0), cities and housing. In 2020, there were as many IoT devices as human beings (around 8 billion), but the annual growth of IoT is ~15% versus 1% for the population (Saulles 2019). In 5 years, the number of IoT devices will have doubled compared to the global population. The increase will not only be quantitative, but also qualitative: the relevance of the IoT in our lives and societies will not stop growing. Unfortunately, cyberattacks targeting the IoT (Kolias *et al.* 2017) will also increase: problems with security and confidentiality remain standard in IoT systems (Alrawi *et al.* 2019).

The term "Internet of Broken Things" reflects this. To change this, research effort on IoT security has increased in recent years (Khan and Salah 2018). MTD (Ghosh *et al.* 2009) is a promising research domain in the area of security that can still help with this effort.

MTD is a cyberdefense paradigm. It tries to balance the asymmetry of information that exists between a system and its attackers. Most systems are static, well-known examples include IP addresses, hardware architecture or the operating system. An attacker faced with a static system potentially has an infinite time to explore it and find weaknesses ("reconnaissance" stage), to prepare and launch and attack. MTD proposes to constantly mutate system components. This will impose constraints on the attacker in the time domain.

There is still a gap between MTD development on classical systems compared to the IoT (Zheng and Namin 2019).

This chapter has two objectives. The first is to present the current state of MTD techniques for the IoT. The second is to contribute to advances in the domain by identifying potential MTD solutions for the IoT and presenting a generic *framework* for them that will allow them to be implemented.

The remainder of this chapter is structured as follows. First of all, we will present general concepts from classical MTD literature in section 11.2 and some related works in section 11.3. In section 11.4, we will summarize MTD techniques for the IoT. We will identify potential MTD solutions for the IoT in section 11.5, and will present our framework proposition in section 11.6. Finally, section 11.7 offers a discussion and avenues for future research, and section 11.8 concludes this chapter.

11.2. Background

The MTD cyberdefense paradigm is based on the principle of building mutating systems. If one vulnerability is exploited, it does not prevent the system from providing its service, and even if it is exploited once, it cannot be exploited again (Chong *et al.* 2009). In this section, we will present a brief chronology and will explore some fundamental concepts for understanding the MTD paradigm.

11.2.1. *Brief chronology of Moving Target Defense*

The concept of modifying a system component to prevent unauthorized parties from disrupting their functioning is not new. Applications of this concept can be found in modern science going back at least a hundred years (Tesla 1903)[1]. In particular, World War II resulted in several advances in this type of application for communication systems (e.g. the Enigma machine, SIGSALY). The context of the World War II also provided the bases of modern cryptography-cryptology thanks to the Colossus computers (Sale 2000). Applications of the concept of defense by constant change in the Internet era can also be found since at least 2001 (Kewley *et al.* 2001). However, it has only been in the course of the last decade that the term "Moving Target Defense" was invented and became a cyberdefense paradigm. Figure 11.1 illustrates the number of scientific publications on the topic of MTD per year.

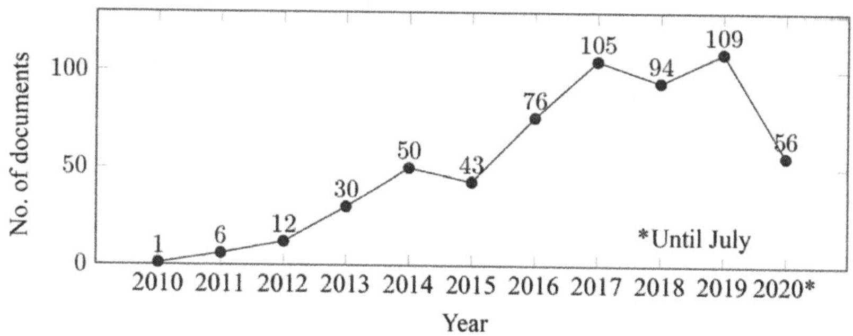

Figure 11.1. *Number of scientific publications per year with "Moving Target Defense" in the title or abstract (before July 2020) (source: Scopus)*

The MTD paradigm was proposed in 2009 in the context of a National Cyberdefense Summit in the United States (Chong *et al.* 2009) with the support of

1 This is a precursor idea of the principles of wireless communication systems with a frequency-hopping spread spectrum.

the federal Networking and Information Technology Research and Development (NITRD) program. Since then, MTD has been an important academic subject in cyberdefense with more than 500 scientific publications up to June 2020. We present a non-exhaustive chronology of MTD milestones in Table 11.1.

2009	First appearance of MTD (Chong *et al.* 2009)
2011	MTD book (Jajodia *et al.* 2011)
2012	Second MTD book (Jajodia *et al.* 2012)
2013	MIT Lincoln Lab survey 1st edition (Okhravi *et al.* 2013)
2014	"Finding Focus on the Blur of MTD" (Okhravi *et al.* 2014) first ACM workshop on MTD (Jajodia et Sun 2014)
2018	MIT Lincoln Lab survey 2nd edition (Ward *et al.* 2018)
2020	7th ACM workshop on MTD

Table 11.1. *Tokens in the MTD domain*

It was not earlier than 2011, with the book by Jajodia *et al.* (2011), that scientific output began to increase. This is the second most cited work in MTD literature, just behind the MTD proposal based on "Software Defined Newtorks" by Jafarian *et al.* (2012). The first survey on MTD propositions by Okhravi *et al.* (2013) has already identified 120 publications. In 2014, Okhravi *et al.* (2014) recognized the heterogenous nature of MTD propositions and laid the foundations of the most used MTD taxonomy until now. Furthermore, the first annual workshop on MTD took place, sponsored by ACM, the workshops were still going in 2020. Finally, returning to Figure 11.1, since 2017 we have observed constant interest in the subject of MTD with around 100 publications per year.

11.2.2. *Fundamental technical and taxonomic principles of MTD*

Literature on MTD can be divided into three domains (Cai *et al.* 2016): theory (Zhuang *et al.* 2015), evaluation (Picek *et al.* 2017) and strategy (Ward *et al.* 2018). The first two address mathematical theory, models and interactions between systems and attackers, and metrics. The MTD strategy domain involves concrete MTD techniques that can be implemented in real systems.

Two important theoretical concepts of MTD systems are the attack surface and the Moving Parameter (MP) or moving target. In short, the attack surface represents the vulnerable components in an IT system. The objective of an MTD system is to dynamize the usable aspects of the system, or, in other words, to "shift" the attack surface. The movement will be concretized with the help of the MP. The MP is a component of the system that will be modified and adapted over the course of time. Thus, choosing and moving an MP that forms part of the attack surface will dynamize it. The MP is one of the most important design elements of an MTD technique. This chapter focuses on MTD techniques, and will be especially applicable to constrained IoT systems.

One broadly accepted taxonomy (Okhravi *et al.* 2014; Ward *et al.* 2018), of existing MTD techniques, and one based on MPs, is the following:

– data: techniques that modify the format, coding or representation of application data;

– software: techniques that dynamically modify an application's code;

– runtime environment: techniques that dynamically modify the execution environment, for example address space layout randomization;

– platform: techniques that modify the properties of the platform, for example the architecture of the processor and the version of the usage system;

– networks: techniques that modify the properties of the operating system, including the protocols or addresses.

In the non-IoT domain, Ward *et al.* (2018) identify 89 different techniques, distributed as follows: data, 9%; software, 15%; execution environment, 35%; platform, 20%; and networks, 21%.

11.3. Related works

11.3.1. *Surveys on MTD techniques*

MTD was proposed at the end of 2009 and the first survey on MTD techniques appeared in 2013 (Okhravi *et al.* 2013). Since then, many studies have been published on MTD techniques (Okhravi *et al.* 2014; Cai *et al.* 2016; Lei *et al.* 2018; Ward *et al.* 2018; Zheng and Namin 2019; Cho *et al.* 2020; Sengupta *et al.* 2020), identifying around 100 distinct techniques.

However, the applicability of MTD for the IoT is not considered in most publications. Only two recent surveys consider MTD for the IoT. The work of Zheng and Namin (2019) has a sub-section on "lightweight MTD"; it identified two techniques and mentions that more MTD techniques are needed for resource-limited devices. The work of Cho *et al.* (2020) has a sub-section on the IoT in the context of a discussion on MTD application domains. It identified four techniques and recognizes that the MTD paradigm seems promising for IoT systems, but with some limits when compared to conventional MTD.

11.3.2. *Frameworks for IoT systems linked to the concept of MTD*

We understand the term *framework* as a set of software or hardware components that serve as the basis for more specific systems or tasks.

There are frameworks for the IoT that can be applied to the MTD paradigm. Husain *et al.* (2013) have proposed a reconfigurable hardware architecture that supports encryption and decryption (REA), based on symmetrical keys. REA aims to use fewer resources and to be faster than AES[2] and to be the hardware basis for techniques based on MTD. Mohsin and Anwar (2016) have proposed a framework based on an ontology for the IoT, aiming to help defense systems against persistent advanced persistent threats.

11.4. LMTD for the IoT: a qualitative survey

In this section, we will review existing MTD techniques for the IoT. We will group the techniques by taxonomy. The aim is to introduce the reader to concrete techniques and emphasize the importance of network category solutions.

11.4.1. *Data: MTD mechanism against side-channel channel attacks based on renegotiating cryptographic keys*

Vuppala *et al.* (2019) have proposed a mechanism based on renegotiating keys in the context of auxiliary channel attacks against embedded systems. Key renegotiation is based on a schema resistant to this type of attack developed by Dziembowski *et al.* (2016). The inputs needed are a long-term key and a random public nonce. In addition to this MTD key, they add noise to the actual message encryption ("masking"). The authors use a (machine learning) method to define an advanced attack and to determine the number of times a key can be used before being compromised. They validate the theoretical lower limit though an implementation on the hardware platform ChipWhisperer using AES. Finally, they suggest using this lower limit as periodicity of their MTD proposal. The schema proposed by Dziembowski *et al.* (2016) has not been implemented.

11.4.2. Software

11.4.2.1. *Diversifying by reconfiguring cryptographic protocols of nodes*

Casola *et al.* (2013, 2014) propose a dynamic and very fine-grained reconfiguration of an IoT node with limited resources. They explain in detail how the proposition can be applied to the cryptographic protocols of the nodes. They describe high-level network messages to initialize valid cryptographic protocol configurations (known beforehand by both parties), and requests and responses from application data to different configurations for the chosen protocol. They are evaluated using

2 Advanced Encryption Standard.

simulation. This consists of an attacker who carries out a "brute force" attack against different encryption schemas. The probability of an attack succeeding depending on the time is calculated for different reconfiguration intervals.

11.4.2.2. *Partitioning and diversification of binary code*

Mahmood and Shila (2016) proposed an MTD technique for the IoT that uses partitioning and diversification of binary code according to context. Partitioning the code according to context means that, for a given context, an IoT node has only stored the minimal code necessary for execution. Diversification of the code is obtained by having different versions of code that fit a given context. The proposition relies on an unconstrained ("online trusted party") that stores the binary code. When the context changes, the IoT node requests the code belonging to the trusted party. The trusted party sends one of the many versions of the code adapted to this context. Finally, the IoT node downloads the new code and deletes the now useless code from the previous context.

11.4.2.3. *μArmor-μScramble: embedded binary security*

Abbasi *et al.* (2019) focused on on binary security for ("deeply embedded") systems. They proposed μArmor as an approach to lessen vulnerabilities linked to memory corruption. The proposal consists of a set of complementary techniques. Among them, μScramble belongs to MTD software. μScramble is a code diversification technique in compilation time. Diversification is obtained by scrambling the registers used, inserting dead code (e.g. NOP operations) and randomizing the function order in the binary code. The authors ran μArmor for an ARM Cortex-M3 based on TI LM3S6965 using OS Zephyr. They evaluated the excess cost of using memory in the most pessimistic scenarios. For μScramble, the overhead is below 5%. They also evaluated the security of μScramble by generating 1,000 variants for several applications, and exposed each application to exploits that are effective in non-diversified versions; on average, the exploits functioned in 10% of variants.

11.4.3. *Runtime environment*

11.4.3.1. *Diversifying the internal interfaces in operating systems*

Researchers from the University of Turku have worked on diversifying interfaces in IoT operating systems (OS).

Hosseinzadeh *et al.* (2015, 2016) first proposed an approach consisting of utilizing diversification and obfuscation techniques for usage systems, APIs and application layer protocols used by IoT devices.

Koivunen *et al.* (2016) concentrated on diversifying the internal interfaces of IoT OSs. They proposed the following interfaces for diversification: system calls, functions in shared libraries, memory space, software shell ("command shell") and network protocols. They evaluated the feasibility of implementing diversification of these interfaces in 10 different IoT usage systems. Unfortunately, most IoT usage systems do not have these interfaces. However, the authors noted that most usage systems have specific internal APIs that lend themselves to diversification. Application layer protocols such as Constrained Application Protocol (CoAP) have also been suggested as appropriate candidates for diversification.

Mäki *et al.* (2016) have designed and implemented diversification at the interfaces of two IoT usage systems, Thingsee and Raspbian. They apply diversification to the link stage of creating executable binaries. They propose mixing the order of the links of the binary objects to introduce diversity into executable binary programs ("memory layout shuffling"), and implemented this proposal in the Thingsee usage system. They measured the effectiveness of this technique against a "Return-Oriented Programming" (ROP) exploit. The diversified Thingsee usage system partially prevented the ROP attack. The exploit stopped the system, but execution of the malicious arbitrary code was prevented. The authors also propose diversifying the symbol names of the shared dynamic libraries ("symbol diversification"), and have implemented this proposal in Raspbian OS as a proof of concept. They obtained an operational usage system, although the functions are limited because of the problems of dynamic linking when executing binaries.

Rauti *et al.* (2018) rely on Koivunen *et al.* (2016) as well as Mäki *et al.* (2016) and examine them more deeply. They studied internal interfaces adapted to diversification in 20 IoT usage systems. They also provided additional technical details on the diversification techniques implemented by Mäki *et al.* (2016).

11.4.3.2. *μArmor-μSSP: embedded binary security*

Abbasi *et al.* (2019) proposed μArmor as an approach for reducing vulnerabilities linked to memory corruption. The proposal consists of a set of additional techniques. Among them, μSSP can be classed as a runtime environment MTD technique. μSSP is based on "stack canaries" techniques and prevent attacks by means of a buffer overflow on the stack memory. A "stack canary" is a random value placed in the memory just before the return pointer and the other metadata on a stack. If a buffer overflow overwrites the metadata of a stack frame, for example to jump to another location, a modification of the "stack canary" can be detected. The effectiveness of μSSP depends on a good source of randomization to generate the "stack canary". To guarantee a good source of randomization , the authors have proposed μRNG, an exclusively software modification of a Cryptographically Secured Pseudo-Random Number Generator (CSPRNG) based on Keccack (Van Herrewege and Verbauwhede 2014). They added a "re-seed" control mechanism adapted to constrained embedded systems. The authors ran μArmor for an ARM Cortex-M3 based on TI LM3S6965 using OS Zephyr. They evaluated the excess cost of using the memory in the most pessimistic scenarios. For μSSP and μRNG the overhead is $<31\%$.

11.4.4. *Platform: diversifying by reconfiguring the IoT node firmware*

Casola *et al.* (2013, 2014) proposed a dynamic and very fine-grained reconfiguration of an IoT node with limited resources. They detail how the proposal can be applied to IoT node firmware. They suppose a wireless sensor network (WSN) formed of nodes of the hardware platform TelosB using TinyOS as a usage system. The proposal to reconfigure the firmware is implemented in TelosB motes. Live (OTA, or Over The Air) dissemination of the complete image of the firmware is not possible in the context of IoT systems with limited resources. Consequently, the solution is based on pre-provisioned local copies of firmware images. The authors have extended the data dissemination protocol Deluge to control the reprogramming parameters of the firmware remotely (e.g. the frequency and type of image). In their evaluation, a node took 6 seconds to reprogram itself with a local storage copy of the firmware. Battista *et al.* (2013) have defined this proposition in more detail and have formalized it in the framework SIREN (SecurIty Reconfiguration for Embedded Nodes). They evaluated the dissemination and reprogramming time of SIREN, the static memory footprint and energy consumption with the help of TelosB motes. Because of the memory constraints, it is not realistic to store more than two firmware in a constrained IoT node.

11.4.5. *Networks*

11.4.5.1. *μMT6D : randomization of IPv6 addresses*

Researchers at Virginia Tech initially proposed MT6D (Dunlop *et al.* 2011) for the randomization of Internet Protocol version 6 (IPv6) addresses in unconstrained networks, and then developed a proposal for constrained IoT systems (Preiss *et al.* 2014; Sherburne *et al.* 2014; Zeitz *et al.* 2017, 2018). MT6D randomizes 64 bits of the interface ID (IID) of IPv6 addresses using a keyless hash function. The transformation function takes as inputs the concatenation of a peripheral device ID, a shared key and the time.

Sherburne *et al.* (2014) first proposed adapting MT6D for IPv6 on Low-Power Wireless Personal Area Networks (6LoWPAN). They discussed some constraints of 6LoWPAN, designed a 6LoWPAN test bench formed of Tmote Sky nodes, and measured the end-to-end delay of HTTP packets between a laptop and a Tmote Sky operating on ContikiOS 2.7.

Preiss *et al.* (2014) have implemented modifications in the IP stack of ContikiOS 2.7 to change the IPv6 addresses periodically. They modified the parameters linked to layer 2 to avoid compression of IPv6 Contiki (Rime) addresses, and the routing protocol stack RPL (IPv6 Routing Protocol for Low-power and Lossy Networks) to propagate the address changes quickly in the network. They added a "sniffer" node to the Sherburne test bench proposal. Using the "sniffer" and information from the border-router RPL node, they evaluated the success of establishing IPv6 routes for address change intervals ranging from 1 to 10 seconds. They obtained a success rate of 99% for all the intervals with the modified RPL stack. If the RPL default parameters are used, the success rate is lower than 15%.

Zeitz *et al.* (2017) focused on the design details and optimizations needed for a fully operational MT6D technique adapted to the IoT, and called it Micro Moving Target IPv6 Defense (μMT6D). They designed two operating modes for μMT6D: a mode in which each IoT node rotates the addresses autonomously, and another in which the RPL *border-router* calculates and propagates the addresses. One optimization proposed is to use lightweight cryptographic hashing algorithms (e.g. PHOTON, Quark or SPONGENT) instead of the secure hashing algorithm SHA-256. They proposed four test benches that mix the real and simulated nodes and networks. They proposed using Cooja from Contiki for the simulated proportion of IoT components, and WiSMotes or Tmote Sky as real IoT nodes. They discussed the need for measures based on additional storage and power, to evaluate the cost of μMT6D and its different configurations.

Later, Zeitz *et al.* (2018) concentrated on the implementation of μMT6D on WiSMotes operating with Contiki 3.0 in Cooja. They used SHA-256 as a hashing algorithm. They evaluated the energy consumption of a node sending one UDP application packet per minute, one energy measurement packet per minute (Collect-View and Energest tools) and rotating the IPv6 address every 5 minutes. They simulate 24 hours and calculate the average hourly electricity consumption. The difference between a μMT6D system and a non-μMT6D system is negligible ($\leq$0.1%). Radio-on (RX) operation represents 99.5% of the power used. If we compare the energy consumption outside RX, the μMT6D system uses around 50% more power.

11.4.5.2. *SARCAST: randomization of IPv6 multicast addresses*

Andrea *et al.* (2017) proposed "Simple Agile RPL multiCAST" (SARCAST), a MTD technique for "multicast". SARCAST aims to reduce Denial of Service (DoS) attacks coming from outside that use multicast IPv6 addresses as an attack vector. The proposal targets multi-hop IoT networks that use the RPL routing protocol. Normalized multicast IPv6 addresses use 112 bits to store a multicast group identifier. SARCAST re-uses these 112 bits in the following way: 24 bits for the group ID, 8 bits to store a sequence number (to reduce replay attacks) and 80 bits to store an agile address. The agile address is a pseudo-random value that serves as a cryptographic validity control. This validity control relies on a keyless hashing algorithm, SHA-1. The most important input parameters of the hashing are: a 32-bit global dynamic counter (broadcast by the RPL root) and a pre-shared 32-bit SALT value. The authors ran SARCAST in Contiki OS and used Cooja to simulate a network of 13 nodes and an attacker. The attacker does not know the secret SALT value. SARCAST effectively reduces multicast DoS attacks from non-adaptive attackers and replay attacks, by a factor between 50% and 90%. The authors validated the results in a real test bench using Zolertia Z1 nodes.

11.4.5.3. *Reconfiguring the topology of IoT Software-Defined Networks (SDN IoT)*

Ge *et al.* (2018) proposed proactive techniques for reconfiguring the topology of IoT SDNs (Software-Defined Networks). First of all, they recognized the feasibility of SDNs for IoT architectures. They assume a multi-hop tree network topology. The attacker'objective is to compromise the base station (gateway). The attacker uses a node as an entry point (the node furthest from the base station).

The authors study two IoT network usage cases. In case 1, the network is formed of patchable and unpatchable nodes. Only patchable nodes can receive firmware updates that can correct vulnerabilities. In case 2, the network is only formed of unpatchable nodes, but some nodes display vulnerabilities that are easy to exploit, whereas others have vulnerabilities that are difficult to exploit. For case 1, the proposal consists of reconfiguring the topology by maximizing the number of reparable nodes along the route to the base station. For case 2, the objective of reconfiguration is similar, but by maximizing the nodes that are difficult to exploit. For both cases, they develop two reconfiguration algorithms to change the topology: an optimal algorithm and a heuristic algorithm. The optimal algorithm does not modify the number of hops from each node in the network. The heuristic method removes the restriction that consists of maintaining the same number of hops and concentrates on optimal local solutions that maximize the number of hops possible from parent nodes.

They use the Akaroa2 discrete-event simulation to validate their proposals. The network is formed of 100 nodes and a base station Ge *et al.* (2018) model the propagation of the attacker according to particular metrics: the probability of the attack succeeding, the impact of the attack, the cost of the attack, average time for compromise, and risk and return on the attacks. They attribute values for the nodes' vulnerabilities using the base score of the Common Vulnerability Scoring System (CVSS). Using a random topology, and for several passages, Akaroa2 calculates the average value of the measures defined. The optimal method for both cases generally increases the effort required of the attacker while maintaining the same number of hops. The heuristic method also generally increases the amount of effort the attacker has to make while maintaining the average length of the shortest path of the network, that is, a performance similar to the optimal.

11.4.5.4. *"Honeypot" MTD for cell-phone assisted IoT systems*

Hamada *et al.* (2018) proposed an MTD technique based on " honeypots" using hospital networks as a case study. An existing hospital network is formed of many connected sub-networks. Each sub-network is endowed with a gateway and connected medical devices that send sensitive information. The proposal relies on a central service (HIoT) operating in the Cloud, and on an application for mobile (H-app) that runs in patients' cell phones. The HIoT orchestrates the MTD strategy. A random sub-set of N cell phones running the H-app will be the chosen application and will be allocated one of the three following roles: true gateway, false gateway and false medical equipment. The topology of the network is going to change. New real gateways will be assigned in the sub-networks with real equipment. False sub-networks will be created, and false medical equipment will be added to real and false sub-networks. After a period of time T, the process is repeated on a new sub-set of H-apps. The authors simulated the rate of change of the topology of a network of unknown size, making between 0 and 30 changes of role per period.

11.4.5.5. *AShA: randomization of MAC-IPv6 addresses*

Nizzi *et al.* (2019) proposed AShA, a shuffling algorithm for IPv6 and MAC addresses on 802.15.4 networks. AShA uses hashing with an HMAC key for pseudo-randomization, and relies on a central coordination node to avoid collisions between addresses. The usage case is that of 6LoWPAN multi-hop networks using the RPL routing protocol. The root of the RPL network plays the role of the central entity. RPL control messages are used to transport the information necessary for AShA to function. Each IoT node can calculate the rotation of addresses using a key, the node's ID and a shared index as HMAC inputs. The shared index is chosen by the coordinator and sent to all the nodes. To avoid address collisions, the coordinator pre-calculates all the node addresses possible (i.e. the HMAC outputs), and sends only a new shared index that will not generate address collisions. The authors have implemented the base function of an AShA server in Python, and have evaluated mathematically the number of node addresses without collision depending on the length of the shared index. The length of the node addresses is 16 bits. Using a shared index of 8 bits in length, a system of 100 nodes without collision is possible. If we use a shared 16-bit index, it is possible to obtain a system without collision for a maximum of 700 nodes.

11.4.6. *Section summary*

In this section, we studied the current state of MTD techniques for the IoT. We reviewed 18 articles corresponding to the 15 different techniques we summarized. The distribution of techniques, according to their taxonomy, is the following: network, 42%; software, 25%; execution environment, 17%; platform, 8%; and data, 8%. Network proposals are predominant, although the total number of techniques is not high enough to allow us to make conclusive statistical statements. In section 11.5, we will concentrate on network techniques.

11.5. Network components in the IoT: a vast domain for MTD

In this section, we will study the network components of IoT systems that may be suitable for MTD techniques.

"Which target to move?" is one of the first questions an MTD system designer must answer. In the previous section, we observed that most solutions for the IoT choose a network parameter. This predominance can be explained by the essential role of the network component in an IoT system. A single IoT node does not have many capacities, but complex services are possible when several are used, paraphrasing the well-known expression: "The system is more than the sum of its nodes." Consequently, cyberattack (Nawir *et al.* 2016) and cyberdefense efforts (Alrawi *et al.* 2019) concentrate on the network component of IoT systems.

In Figure 11.2, we can observe the MTD taxonomies while concentrating on the network component with the help of OSI layer model. Below, we will explore each layer with concrete examples of the components of IoT systems that can be used for MTD.

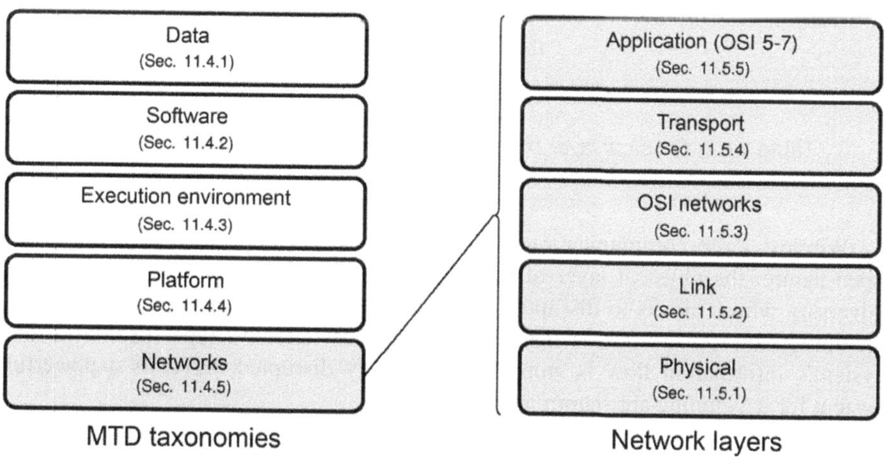

Figure 11.2. *MTD taxonomies and network layers (with a reference to the section in which we study them. For a color version of this figure, see www.iste.co.uk/chalouf/intelligent.zip)*

11.5.1. *Physical layer*

In Table 11.2, we can observe some common components of the physical layer and its variants.

Component	Variants
Transmission medium	Radio waves Transmission lines (wired) its (atmosphere) Optical wireless communication (visible, IR, UV) Molecular communication (for example pheromones, ethanol)
Radio modulation	PSK, FSK, ASK, QAM
Carrier frequency	868 MHz, 2,4 GHz, 5,8 GHz, 3 kHz (VLF, non-ISM), 12-18 GHz (Ku-band, non-ISM)
Multiplexing	Time; frequency: OFDM Spread spectrum techniques (DSSS, FHSS) Code divison or code multi-plexing (CDMA) *Time Slotted Channel Hopping* (IEEE TSCH)
Physical topology	Mesh, star, bus, point-to-point, ring/tree
Commercial technology	IEEE 802.15.4 LPWAN (SubGHz): LoRa, SigFOX, NB-IoT LTE-Cat0/M IEEE 802.11ah (SubGHz)/802.11ax (2.4GHz) Bluetooth (2.4 GHz)

Table 11.2. *Physical layer components (OSI L1) and variants for the IoT*

Wireless radio communication is a key technology for the IoT. Because of its open nature, the physical layer of wireless systems is a high priority target for an adversary whose aim is to disrupt the normal behavior of the system. In particular, "jamming attacks" are one of the simplest and most effective types of attack: the system's information flow is stopped or seriously disrupted. MTD is a powerful system for developing anti-jamming techniques.

In general, the presence of several radio technologies with the same node forms the exception. Nevertheless, more and more devices are available that support SubGHz technology (long range) and a 2.4 GHz technology. In the same frequency range, spread spectrum anti-jamming techniques seem promising. Radio technology defined by software "Software-Defined Radio", in combination with MTD, could change the deal if they become economically accessible.

11.5.2. *Link layer*

In Table 11.3, we can observe some common components of the link layer and its variants.

Component	Variants
MEdia access control Medium Access Control (MAC)	Collision recovery: ALOHA, R-ALOHA Collision avoidance: CSMA/CA, MACAW Without collision: MS-ALOHA (CDMA, OFDMA)
L2 routing ("Mesh-Under")	LOAD, HWMP/IEEE 802.11s, STP, B.A.T.M.A.N.
Metrics for L2 routing	ETX, RSSI, Fuzzy LQE, Four Bit
Frame format	802.15.4, LoraWAN, Profinet, Ethernet, générique
L2 address	EUI-64, MAC-48
L2 encryption	In general, L2 frame standards define their own secure format

Table 11.3. *Components of the link layer (OSI L2) and its variants for the IoT*

Randomization of 12 L2 addresses has already been studied in the context of MTD for the IoT (Nizzi *et al.* 2019). Information security at the level of the link layer is also increasingly prevalent in standards (and use cases) and includes many parameters that can be modified. Among the other components mentioned, the parameters of a MAC method or metrics for L2 routing could be suitable for applying an MTD.

11.5.3. *OSI network layer*

The network layer of the OSI model (and of the TCP/IP model) has been the most explored by attackers and existing MTD techniques. The IP protocol is at the center of this layer. It allows remote elements to interact with the IoT network. This connectivity is essential to providing new services. Unfortunately, this is also a point of entry for attackers, who no longer need to be physically close to the system. In Table 11.4, we can observe some shared components of this layer and its variants.

In addition to the IP protocol, two other protocols are facilitators of the IoT: the RPL routing protocol and the IPv6 6LoWPAN header compression protocol. As for existing attacks, the work of Pongle and Chavan (2015) concentrates uniquely on attacks targeting these two protocols. In more general literature on attacks on IoT systems (Nawir *et al.* 2016; Deogirikar and Vidhate 2017), OSI network layer attacks represent the majority of the cases studied.

Three of the MTD techniques studied in section 11.4.5 target the IP protocol. One of them also targets the RPL protocol. Finally, one of them addresses the network's logical topology relying on the SDN paradigm, although in simulated fashion.

Component	Variants
Logical topology	"Mesh" (meshed), ⋆, hybrid
L3 routing ("Route-Over")	RPL (RFC 6550), LOADng, BABEL (RFC 6126), LEACH/MR-LEACH, AODV, DSE, OLSR
L3 routing – metrics	ETX, RSSI, Fuzzy LQE, Four Bit
L3 routing – RPL parameters	*Objective Functions* (OF): OF0, MRHOF, new OF (ddifferent OFs may lead to different topologies)
IPv6 compression (6LoWPAN)	No compression, 6LoWPAN (RFC 4944), SCHC (RFC 8724)
IP address	Static/EUI-64, DHCP, *Micro Moving Target IPv6 Defense* (μMT6D), SARCAST (*IPv6 Multicast Address MTD*)
L3 security (IPsec)	IKEv2, ESP (8 variants), ESP and AH (8 variants)

Table 11.4. *Components of the network layer (OSI L3) and its variants for the IoT*

We consider that this layer still has great potential for MTD techniques. Security techniques that target the RPL protocol in particular have been moderately explored. The SDN paradigm is also very promising. However, most SDN solutions are not adapted to the constraints of the IoT. Research effort is needed to make SDN for the IoT a reality. As for the randomization of IP addresses, this is the most mature MTD technique for the IoT and it can plausibly (Zeitz *et al.* 2018) be used in real deployments.

11.5.4. *Transport layer*

In Table 11.5, we can observe some common components of the transport layer and its variants.

Component	Variants
Transport protocol	UDP, TCP, SCTP
Transport protocol (ports)	Known ports, port randomization, *port-knocking* (specific sequences or authenticated packets)
Transport security	TLS, DTLS, Compact TLS (cTLS) 1.3.
"Cipher suite" (D)TLS	Combination of authentication algorithms and encryption Choice of Cipher suite: more than 50 exist

Table 11.5. *Components of the transport layer (OSI L4) and its variants for the IoT*

UDP is the transport protocol of choice for the most constrained IoT nodes. Security services for this layer, such as DTLS or more recently cTLS, are also used in real deployments. However, security alternatives at the application layer are currently emerging. As for MTD solutions, we will propose a UDP port-hopping technique in section 11.6.2. Parameters linked to DTLS (such as "cipher suites" and the associated keys) could be suitable for implementing MTD techniques.

11.5.5. *Application layer*

Finally, in Table 11.6, we can observe some common components of the application layer of the TCP/IP (OSI 5-7) model and its variants. This layer could include some solutions resulting from the "data" taxonomy of MTD. In general, this layer processes distributed protocols or application data that will be transported from one node to another, but some proposals could also be implemented in non-distributed systems and classed as "data" techniques.

Component	Variants
Application protocol	CoAP (RFC 7252), MQTT, LWM2M, HTTP, XMPP
Media-Type	CBOR, JSON, Binary, XML, HTML/javascript/jpg
Resource locators RESTful interfaces	Resource discovery, well-known resources, resource directories
Application security	COSE (8152) , OSCORE (RFC 8613), JOSE (RFC 7515)
Application security: COSE	Algorithm change, key change, cascade encryption
ACE-OAuth Framework for the IoT (IETF)	"proof-of-possession" tokens, "bearer" tokens Many components of tokens can be modified: expiry, associated authorizations, associated key
Application plain text (syntax)	We can modify the syntax while still keeping the semantics

Table 11.6. *Application (OSI L5-7) layer components and variants for the IoT*

Application layer components are very numerous in the IoT. The most important is the RESTful CoAP protocol. In terms of protocols linked to security, application layer solutions emerge as COSE (RFC 8152) and OSCORE (RFC 8613). They offer great flexibility since the security parameters can be configured per message (unlike the DTLS transport layer where they are configured per session). MTD techniques could exploit this flexibility per message. ACE-OAuth[3] is an authorization framework based on OAuth 2.0. The use of authorization "tokens" is at the heart of the framework, and these tokens have many parameters also suitable for MTD.

Nevertheless, no MTD techniques use this. This layer has great potential to be used as a source of new MTD techniques.

11.5.6. *Section summary*

In this section, we have reviewed the common network components of IoT systems by separating them by the layers of the OSI model. We have identified the relevant components and their variants. These components are potentially useful for implementing MTD techniques if enough variants can be used realistically.

The use of components such as MAC and IP addresses has already been explored in MTD literature for the IoT. However, we have identified many other components that are promising for MTD and which have not been explored in depth, such as application layer components.

In section 11.6, we develop a generic MTD framework that can shift a distributed and arbitrary "moving parameter" in an IoT system.

11.6. An MTD *framework* for the IoT

In this section, we present a generic framework that can be used to instantiate MTD strategies adapted to the IoT. We have called this framework IANVS (pronounced *Janus*) in honor of the Roman god of transition.

We have proposed this framework more exhaustively in a previous work (Navas *et al.* 2020). We include it briefly in this chapter, since it serves as a tool for building MTD techniques on the components identified in section 11.5. As an example, we instantiate it with the UDP port as a moving target.

11.6.1. *Proposition: components*

IANVS abstracts, generalizes and links the common components of distributed MTD strategies. One concrete MTD technique can use IANVS as an archetype. The main objective of IANVS is to facilitate future work on the design and implementation of MTD techniques for the IoT. Figure 11.3 illustrates IANVS and its components. Below, we detail each component.

---

3 Authentication and Authorization for Constrained Environments (ACE) using the OAuth 2.0 Framework (ACE-OAuth).

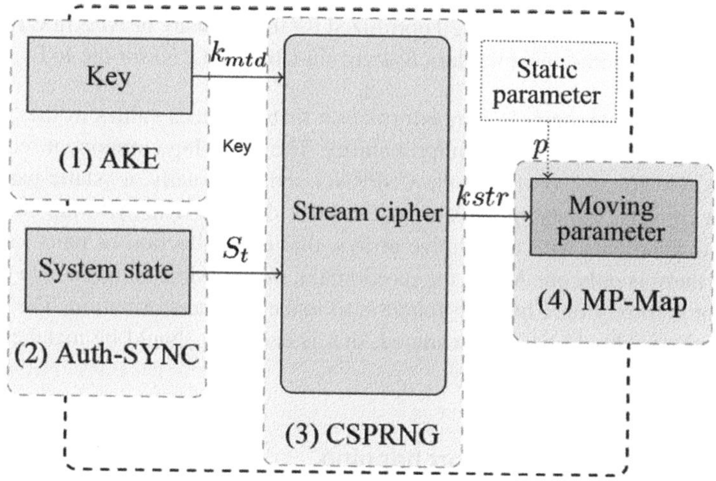

Figure 11.3. *Components of the proposed MTD framework. For a color version of this figure, see www.iste.co.uk/chalouf/intelligent.zip*

1) AKE: a mechanism for establishing authenticated keys ("Authenticated Key Establishment"). The AKE component provides a cryptographic (k_{mtd}) key. The k_{mtd} key is secret and should only be shared by parties trusted by an MTD strategy. AKE is linked to the problem of "bootstrapping" secure keys. Bootstrapping of keys is a difficult problem to solve and, in general, it relies on pre-shared cryptographic material or on a "Trusted Third Party". Candidate AKE components in the context of the constrained IoT are AKE protocols such as Ephemeral Diffie-Hellman Over COSE (EDHOC) (Selander *et al.* 2020), our previous AKE work based on "nonces" (Navas *et al.* 2016), or a DTLS "handshake" DTLS with a lightweight "cipher suite".

2) Auth-SYNC: authenticated state synchronization mechanism. The Auth-SYNC component provides a system state value S_t. The value S_t should be authenticated and recent, but not necessarily secret. Auth-SYNC is linked to the problem of synchronizing the secure time. Currently, no solutions exist that are appropriate to the IoT. We evaluate the state of the art and make a proposal in our previous works (Navas and Toutain 2018).

3) CSPRNG: a cryptographically secure pseudo-random number generator, using a cryptographic construction family of "stream ciphers". This component lies at the heart of the framework. It requires two inputs: (1) a cryptographic key (k_{mtd}) and (2) an authenticated system state (S_t). They are provided by the AKE and Auth-SYNC components, respectively. The CSPRNG produces an output, a stream of binary keys, or a cryptographically strong "keystream" ($kstr$). A stream cipher (Banik *et al.* 2018) should be used, such as CSPRNG. There are many options adapted to the IoT (Philip

et al. 2017), for example, ChaCha20 optimized for the software or AES in CTR mode, which is integrated into most modern System-on-Chips (SOCs) for the IoT.

4) The final component is a transformation that produces values in the "Moving Parameter" (MP) domain with equiprobability. The MP-Map component requires as inputs the "keystream" ($kstr$) of the CSPRNG and, optionally, a "static parameter" value p. It converts inputs into a value in the MP's domain. The MP-Map component is linked to problems with associative arrays, that is a collection of pairs *key-value*. Although there is only one MP in the (co)domain, one possible solution is to use hash tables, that is to rely on a hashing function to make the transformation. The uniform hash hypothesis should be approximated, that is each *key* should be matched with a domain *value* with equiprobability.

11.6.2. *Instantiation:* UDP port hopping

In this section, we propose a concrete MTD technique. This strategy corresponds to the "network" category of MTD strategies. The MP in this strategy is a service's UDP port number.

The pseudo-randomization of TCP and UDP port numbers has already been proposed in literature not linked to the IoT (Lee and Thing 2004), and is known by the name of port hopping. Well-known port numbers are needed to discover and use network services. However, the static nature enables DoS attacks (Liang *et al.* 2016) (e.g. by "flooding" from a well-known port). They are also the entry point for adversaries in the "reconnaissance" stage of more sophisticated attacks that target the upper layers. Port-hopping techniques can effectively reduce this type of attack. We will use IANVS to instantiate a solution of this type.

To instantiate IANVS, its four components should be clearly defined. For this technique, the components are defined as follows:

– AKE: a pre-shared symmetrical 128-bit key;

– Auth-SYNC: we use periodic changes of MP, and the Network Time Protocol (NTP)[4] to synchronize the clock in real time (RTC) of the distributed MTD nodes;

– CSPRNG: we use ChaCha20 with 20 rounds. The two inputs are the AKE's unmodified 128-bit key and a 64-bit nonce derived from the NTP time;

– MP-Map: UDP port numbers go from 0 to 65,535 (16 bits). The cardinality of the MP domain is therefore $|MP| \leq |2^{16}|$. We propose a strategy for a single port hop. If the port hop is the only UDP port open, it is easy to use the 16 bits for the hopping port ($|MP| = |2^{16}|$). If p is the well-known port number to transform, we

4 "Network Time Security for the NTP" (IETF draft) should be used in a real deployment.

apply a xor bit per bit with the first 16 bits of the output of the ChaCha20 $kstr$. This transformation is equivalent to using ChaCha20 stream cipher to encrypt p. The port hop p_{mtd} is equal to $p \oplus kstr_{0..15}$.

We have implemented and tested this solution in real hardware. We refer the reader to the extended work (Navas *et al.* 2020) to find the results and analysis, and to the reference (Sandaker and Navas 2020) for the implementation source code and the raw results[5]. To summarize, the theoretical probability of a brute force attacker finding the right port can be expressed with the binomial distribution. The exact parameters of the distribution will depend on the frequency of movement of the MTD technique (compared to the frequency of attacks) and on the number of ports available for the jump. The experimental results have validated the theory.

11.7. Discussion and avenues for future research

MTD for traditional systems is a recent but developed cyberdefense paradigm, already with more than 80 distinct techniques. However, its application in the IoT domain is still not as mature.

In sections 11.1–11.4, we concentrated on understanding the MTD paradigm by presenting a brief chronology with milestones and general concepts. In the same line of investigation, we have reviewed the current state of MTD techniques for the IoT, by concentrating on particular techniques. There are currently 15 different techniques, and we have observed that many of these (42%) form part of the MTD "network" category. This predominance is not surprising, since the network component plays a crucial role in IoT systems.

For sections 11.5 and 11.6, we wanted to propose a constructive approach to the domain. We believe that MTD has great potential for improving the resilience of the IoT. We concentrated on the network component of IoT systems. We intended to make explicit the particular domains where it could be applied, and we chose an approach using the network layers of the OSI model. At each layer, we identified several components adapted to MTD technology, and we have briefly tackled the most promising. We have also shown that most existing solutions concentrate on the randomization of IP or MAC addresses. Finally, we have presented a generic framework that can be used to instantiate concrete and distributed MTD techniques adapted to the IoT. We have provided an example of implementing a UDP port-hopping solution. We have used a modular approach and shared the source code. With this framework approach, we wanted to show the feasibility of new MTD techniques for the IoT and to simplify the job of future designers of MTD techniques.

---

5 We also refer the reader to https://github.com/ renzoe/IANVS/tree/v1.0.0.

As future axes for research, "Software-Defined" technologies, such as radio (SDR) or network (SDN), are particularly promising, or will be as soon as they become economically or technically possible for the IoT. They will permit the implementation of MTD techniques at the level of the physical layer and the logical topology, respectively, without it being necessary to redeploy nodes. The RPL routing protocol used in meshed IoT networks can also gain robustness thanks to the implementation of mechanisms inspired by the MTD paradigm. We have also shown that the "application" layer is almost unexplored by MTD techniques, but that it is an excellent source of parameters suitable for the MTD.

11.8. Conclusion

In this chapter, we have presented the MTD paradigm in the context of the IoT. The MTD randomizes systems' components and mitigates the effect of otherwise trivial attacks. We studied the current state of MTD techniques for the IoT. Fifteen distinct techniques exist, and most of them concentrate on the network component of IoT systems. In addition, they focus on randomizing IP or MAC addresses. With a constructive approach, we have examined the network layers of current IoT systems by identifying the components that could be suitable for the MTD paradigm. We have identified several unexplored domains. Finally, we have proposed a generic *framework* that can be used to instantiate particular MTD techniques, and we have used it to instantiate a particular technique. We hope to have raised interest in the MTD paradigm as a viable method of improving the security of IoT systems.

11.9. References

Abbasi, A.,Wetzels, J., Holz, T., Etalle, S. (2019). Challenges in designing exploit mitigations for deeply embedded systems. In *2019 IEEE European Symposium on Security and Privacy (EuroS&P)*. IEEE, Stockholm.

Alrawi, O., Lever, C., Antonakakis, M., Monrose, F. (2019). SoK: Security evaluation of home-based IoT deployments. In *2019 IEEE Symposium on Security and Privacy*. IEEE, San Francisco.

Andrea, K., Gumusalan, A., Simon, R., Harney, H. (2017). The design and implementation of a multicast address moving target defensive system for Internet-of-Things applications. In *MILCOM 2017 – IEEE Military Communications Conference*. IEEE, Baltimore.

Banik, S., Mikhalev, V., Armknecht, F., Isobe, T., Meier, W., Bogdanov, A., Watanabe, Y., Regazzoni, F. (2018). Towards low energy stream ciphers. *IACR Transactions on Symmetric Cryptology*, 2, 1–19.

Battista, E., Casola, V., Mazzeo, A., Mazzocca, N. (2013). Siren: A feasible moving target defence framework for securing resource-constrained embedded nodes. *International Journal of Critical Computer-Based Systems*, 4(4), 374–392.

Cai, G.-L., Wang, B.-S., Hu, W., Wang, T.-Z. (2016). Moving target defense: state of the art and characteristics. *Frontiers of Information Technology & Electronic Engineering*, 17(11), 1122–1153.

Casola, V., De Benedictis, A., Albanese, M. (2013). A moving target defense approach for protecting resource-constrained distributed devices. In *2013 IEEE 14th International Conference on Information Reuse & Integration (IRI)*. IEEE, San Francisco.

Casola, V., De Benedictis, A., Albanese, M. (2014). A multi-layer moving target defense approach for protecting resource-constrained distributed devices. In *Integration of Reusable Systems*, Bouabana-Tebibel, T., Rubin, S.H. (eds). Springer, Cham.

Cho, J.-H., Sharma, D.P., Alavizadeh, H., Yoon, S., Ben-Asher, N., Moore, T.J., Kim, D.S., Lim, H., Nelson, F.F. (2020). Toward proactive, adaptive defense: A survey on moving target defense. *IEEE Communications Surveys & Tutorials*, 22(1), 709–745.

Chong, F., Lee, R., Acquisti, A., Horne, W., Palmer, C., Ghosh, A., Pendarakis, D., Sanders, W., Fleischman, E., Teufel III, H. *et al.* (2009). National cyber leap year summit 2009: Co-chairs' report, NITRD Program [Online]. Available at: https://www.nitrd.gov/nitrdgroups/images/b/bd/National_Cyber_Leap_Year_Summit_2009_CoChairs_Report.pdf [Accessed 22 June 2020].

Deogirikar, J. and Vidhate, A. (2017). Security attacks in IoT: A survey. In *2017 International Conference on I-SMAC (IoT in Social, Mobile, Analytics and Cloud) (I-SMAC)*. IEEE, Palladam.

Dunlop, M., Groat, S., Urbanski, W., Marchany, R., Tront, J. (2011). Mt6d: A moving target IPv6 defense. In *MILCOM 2011 – Military Communications Conference*. IEEE, Baltimore.

Dziembowski, S., Faust, S., Herold, G., Journault, A., Masny, D., Standaert, F.-X. (2016). Towards sound fresh re-keying with hard (physical) learning problems. In *Annual International Cryptology Conference*. IACR, Santa Barbara.

Ge, M., Hong, J.B., Yusuf, S.E., Kim, D.S. (2018). Proactive defense mechanisms for the software-defined Internet of Things with non-patchable vulnerabilities. *Future Generation Computer Systems*, 78, 568–582.

Ghosh, A.K., Pendarakis, D., Sanders, W.H. (2009). Moving target defense co-chair's report: National Cyber Leap Year Summit 2009. Technical Report, Federal Networking and Information Technology Research and Development (NITRD) Program, Arlington.

Hamada, A.O., Azab, M., Mokhtar, A. (2018). Honeypot-like moving-target defense for secure IoT operation. In *2018 IEEE 9th Annual Information Technology, Electronics and Mobile Communication Conference*. IEEE, Vancouver.

Hosseinzadeh, S., Rauti, S., Hyrynsalmi, S., Leppänen, V. (2015). Security in the Internet of Things through obfuscation and diversification. In *2015 International Conference on Computing, Communication and Security*. IEEE, Pointe-aux-Piments.

Hosseinzadeh, S., Hyrynsalmi, S., Leppänen, V. (2016). Obfuscation and diversification for securing the Internet of Things (IoT). In *Internet of Things: Principles and Paradigms*, Buyya, R., Dastjerdi, A.V. (eds). Elsevier, Amsterdam.

Husain, M.I., Courtright, K., Sridhar, R. (2013). Lightweight reconfigurable encryption architecture for moving target defense. In *MILCOM 2013*. IEEE, San Diego.

Jafarian, J.H., Al-Shaer, E., Duan, Q. (2012). Openflow random host mutation: Transparent moving target defense using software defined networking. In *1st Workshop on Hot Topics in Software Defined Networks*. SIGCOMM, Helsinki.

Jajodia, S. and Sun, K. (2014). MTD 2014: First ACM workshop on moving target defense. In *2014 ACM SIGSAC Conference on Computer and Communications Security*. ACM, New York. doi.acm.org/10.1145/2660267.2660385.

Jajodia, S., Ghosh, A.K., Swarup, V., Wang, C., Wang, X.S. (2011). *Moving Target Defense: Creating Asymmetric Uncertainty for Cyber Threats*. Springer-Verlag, New York.

Jajodia, S., Ghosh, A.K., Subrahmanian, V.S., Swarup, V., Wang, C., Wang, X.S. (2012). *Moving Target Defense II: Application of Game Theory and Adversarial Modeling*. Springer-Verlag, New York.

Kewley, D., Fink, R., Lowry, J., Dean, M. (2001). Dynamic approaches to thwart adversary intelligence gathering. In *DARPA Information Survivability Conference and Exposition II*. IEEE, Anaheim.

Khan, M.A. and Salah, K. (2018). IoT security: Review, blockchain solutions, and open challenges. *Future Generation Computer Systems*, 82, 395–411.

Koivunen, L., Rauti, S., Leppänen, V. (2016). Applying internal interface diversification to IoT operating systems. In *2016 International Conference on Software Security and Assurance*. IEEE, Sankt Pölten.

Kolias, C., Kambourakis, G., Stavrou, A., Voas, J. (2017). DDos in the IoT: Mirai and other botnets. *Computer*, 50(7), 80–84.

Lee, H.C. and Thing, V.L. (2004). Port hopping for resilient networks. In *IEEE 60th Vehicular Technology Conference*. IEEE, Los Angeles.

Lei, C., Zhang, H.-Q., Tan, J.-L., Zhang, Y.-C., Liu, X.-H. (2018). Moving target defense techniques: A survey. *Security and Communication Networks* [Online]. Available at: https://www.hindawi.com/journals/scn/2018/3759626/.

Liang, L., Zheng, K., Sheng, Q., Huang, X. (2016). A denial of service attack method for an IoT system. In *8th International Conference on Information Technology in Medicine and Education*. IEEE, Fuzhou.

Mahmood, K. and Shila, D.M. (2016). Moving target defense for Internet of Things using context aware code partitioning and code diversification. In *2016 IEEE 3rd World Forum on the Internet of Things (WF-IoT)*. IEEE, Reston.

Mäki, P., Rauti, S., Hosseinzadeh, S., Koivunen, L., Leppänen, V. (2016). Interface diversification in IoT operating systems. In *9th International Conference on Utility and Cloud Computing*. ACM, Shanghai.

Mohsin, M. and Anwar, Z. (2016). Where to kill the cyber kill-chain: An ontology-driven framework for IoT security analytics. In *2016 International Conference on Frontiers of Information Technology*. IEEE, Islamabad.

Navas, R.E. and Toutain, L. (2018). Late: A lightweight authenticated time synchronization protocol for IoT. In *2018 Global Internet of Things Summit (GIoTS)*. IEEE, Bilbao.

Navas, R.E., Lagos, M., Toutain, L., Vijayasankar, K. (2016). Nonce-based authenticated key establishment over OAuth 2.0 IoT proof-of-possession architecture. In *IEEE 3rd World Forum on Internet of Things (WF-IoT)*. IEEE, Reston.

Navas, R.E., Sandaker, H., Cuppens, F., Cuppens, N., Toutain, L., Papadopoulos, G. (2020). IANVS: A moving target defense framework for a resilient Internet of Things. In *25th IEEE Symposium on Computers and Communications (ISCC)*. IEEE, Rennes.

Nawir, M., Amir, A., Yaakob, N., Lynn, O.B. (2016). Internet of Things (IoT): Taxonomy of security attacks. In *2016 3rd International Conference on Electronic Design (ICED)*. IEEE, Phuket.

Nizzi, F., Pecorella, T., Esposito, F., Pierucci, L., Fantacci, R. (2019). IoT security via address shuffling: The easy way. *IEEE Internet of Things Journal*, 6(2), 3764–3774.

Okhravi, H., Rabe, M.A., Mayberry, T.J., Leonard, W.G., Hobson, T.R., Bigelow, D., Streilein, W.W. (2013). Survey of cyber moving target techniques. Technical report, MIT Lincoln Laboratory, Lexington.

Okhravi, H., Hobson, T., Bigelow, D., Streilein, W. (2014). Finding focus in the blur of moving-target techniques. *IEEE Security & Privacy*, 12(2), 16–26.

Philip, M.A. and Vaithiyanathan (2017). A survey on lightweight ciphers for IoT devices. In *2017 International Conference on Technological Advancements in Power and Energy (TAP Energy)*. IEEE, Vallikavu.

Picek, S., Hemberg, E., O'Reilly, U.-M. (2017). If you can't measure it, you can't improve it: Moving target defense metrics. In *2017 Workshop on Moving Target Defense*. ACM, Dallas.

Pongle, P. and Chavan, G. (2015). A survey: Attacks on RPL and 6LoWPAN in IoT. In *2015 International Conference on Pervasive Computing (ICPC)*. IEEE, Pune.

Preiss, T., Sherburne, M., Marchany, R., Tront, J. (2014). Implementing dynamic address changes in contikios. In *International Conference on Information Society*. IEEE, London.

Rauti, S., Koivunen, L., Mäki, P., Hosseinzadeh, S., Laurén, S., Holvitie, J., Leppänen, V. (2018). Internal interface diversification as a security measure in sensor networks. *Journal of Sensor and Actuator Networks*, 7(1), 12.

Sale, A.E. (2000). Lorenz and colossus [military cryptography]. In *13th IEEE Computer Security Foundations Workshop*. IEEE, Cambridge.

Sandaker, H. and Navas, R.E. (2020). IANVS: An MTD framework for a resilient IoT. Porthopping code and data. doi:10.5281/zenodo.3839686.

Saulles, M.D. (2019). IoT statistics – Information matters [Online]. Available at: https://informationmatters.net/internet-of-things-statistics/ [Accessed 22 June 2020].

Selander, G., Mattsson, J.P., Palombini, F. (2020). Ephemeral Diffie-Hellman Over COSE (EDHOC). Internet-draft draft-selander-lake-edhoc-01, Internet Engineering Task Force [Online]. Available at: https://datatracker.ietf.org/doc/html/draft-selander-lake-edhoc-01 [Accessed 22 June 2020].

Sengupta, S., Chowdhary, A., Sabur, A., Alshamrani, A., Huang, D., Kambhampati, S. (2020). A survey of moving target defenses for network security. *IEEE Communications Surveys & Tutorials*, 22(3), 1909–1941.

Sherburne, M., Marchany, R., Tront, J. (2014). Implementing moving target IPv6 defense to secure 6LoWPAN in the Internet of Things and smart grid. In *9th Annual Cyber and Information Security Research Conference*. ACM, Oak Ridge.

Tesla, N. (1903). System of signaling. US Patent 725, 605.

Van Herrewege, A. and Verbauwhede, I. (2014). Software only, extremely compact, keccak-based secure PRNG on ARM Cortex-M. In *2014 51st ACM/EDAC/IEEE Design Automation Conference (DAC)*. IEEE, San Francisco.

Vuppala, S., Mady, A.E., Kuenzi, A. (2019). Moving target defense mechanism for sidechannel attacks. *IEEE Systems Journal*, 14(2), 1–10.

Ward, B.C., Gomez, S.R., Skowyra, R., Bigelow, D., Martin, J., Landry, J., Okhravi, H. (2018). Survey of cyber moving targets second edition. Technical report, MIT Lincoln Laboratory, Lexington.

Zeitz, K., Cantrell, M., Marchany, R., Tront, J. (2017). Designing a micro-moving target IPv6 defense for the Internet of Things. In *2017 IEEE/ACM Second International Conference on Internet-of-Things Design and Implementation (IoTDI)*. IEEE, Pittsburgh.

Zeitz, K., Cantrell, M., Marchany, R., Tront, J. (2018). Changing the game: A micro moving target IPv6 defense for the Internet of Things. *IEEE Wireless Communications Letters*, 7(4), 578–581.

Zheng, J. and Namin, A.S. (2019). A survey on the moving target defense strategies: An architectural perspective. *Journal of Computer Science and Technology*, 34(1), 207–233.

Zhuang, R., Bardas, A.G., DeLoach, S.A., Ou, X. (2015). A theory of cyber attacks: A step towards analyzing MTD systems. In *2nd ACM Workshop on Moving Target Defense*. ACM, Denver.

List of Authors

Soraya AIT-CHELLOUCHE
Inria
CNRS
IRISA
University of Rennes 1
France

Badr BENMAMMAR
LTT
University Abou Bekr Belkaid Tlemcen
Algeria

Olivier BERDER
IRISA
CNRS
University of Rennes 1
Lannion
France

Sinda BOUSSEN
Mediatron
University of Carthage
Tunis
Tunisia

Mohamed-Aymen CHALOUF
IRISA
University of Rennes 1
Lannion
France

Matthieu GAUTIER
IRISA
CNRS
University of Rennes 1
Lannion
France

Yassine HADJADJ-AOUL
Inria
CNRS
IRISA
University of Rennes 1
France

Omessaad HAMDI
IEEE
IMT Atlantique
Rennes
France

Ahmad KHALIL
LIB
Bourgogne Franche-Comté University
Dijon
France

Francine KRIEF
LaBRI
Bordeaux INP
Talence
France

Guillaume LE GALL
IRISA
IMT Atlantique
Rennes
France

Nader MBAREK
LIB
Bourgogne Franche-Comté University
Dijon
France

Léo MENDIBOURE
LaBRI
University of Bordeaux
Talence
France

Nicolas MONTAVONT
IRISA
IMT Atlantique
Rennes
France

Renzo E. NAVAS
IRISA
IMT Atlantique
Rennes
France

Georgios Z. PAPADOPOULOS
IRISA
IMT Atlantique
Rennes
France

Amal SAMMOUD
SYSCOM
ENIT
University of Tunis El Manar
Tunisia

Tidiane SYLLA
ISA
USTTB
Bamako
Mali

Olivier TOGNI
LIB
Bourgogne Franche-Comté University
Dijon
France

Laurent TOUTAIN
IRISA
IMT Atlantique
Rennes
France

Index

Printed and bound by CPI Group (UK) Ltd, Croydon, CR0 4YY
13/09/2026
14960996-0004